127 Advances in Polymer Science

QD
281
P6
F66
V.127
CHEM

Springer
*Berlin
Heidelberg
New York
Barcelona
Budapest
Hong Kong
London
Milan
Paris
Santa Clara
Singapore
Tokyo*

Polymer Synthesis/ Polymer Catalysis

With contributions by
B. Améduri, M. Arndt, B. Boutevin, T. Endo,
Ph. Gramain, W. Kaminsky, E. Ruckenstein, Y. Yagci

With 63 Figures

 Springer

This series presents critical reviews of the present position and future trends in modern polymer research. It is addressed to all polymer and material scientists in industry and the academic community who wish to keep abreast of advances in the topics covered.

As a rule, contributions are specially commissioned. The editors and publishers will, however, always be pleased to receive suggestions and supplementary information. Papers are accepted for "Advances in Polymer Science" in English.

In references "Advances in Polymer Science" is abbreviated Adv. Polym. Sci. and is cited as a journal.

Springer WWW homepage: http://www.springer.de

ISBN 3-540-61288-2 Springer-Verlag Berlin Heidelberg NewYork
ISBN 0-387-61288-2 Springer-Verlag NewYork Berlin Heidelberg

This work is subject to copyright. All rights are reserved, whether the whole or part of the material is concerned, specifically the rights of translation, reprinting, re-use of illustrations, recitation, broadcasting, reproduction on microfilms or in other ways, and storage in data banks. Duplication of this publication or parts thereof is only permitted under the provisions of the German Copyright Law of September 9, 1965, in its current version, and a copyright fee must always be paid.

© Springer-Verlag Berlin Heidelberg New York 1997
ISSN 0065-3195
Printed in Germany

The use of registered names, trademarks, etc. in this publication does not imply, even in the absence of a specific statement, that such names are exempt from the relevant protective laws and regulations and therefore free for general use.

Typesetting: Macmillan India Ltd., Bangalore-25
SPIN: 10508872   02/3020 - 5 4 3 2 1 0 - Printed on acid-free paper

Editors

Prof. Akihiro Abe, Department of Industrial Chemistry, Tokyo Institute of Polytechnics, 1583 Iiyama, Atsugi 243-02, Japan

Prof. Henri Benoit, CNRS, Centre de Recherches sur les Macromolécules, 6, Rue Boussingault, 67083 Strasbourg Cedex, France

Prof. Hans-Joachim Cantow, Freiburger Materialforschungszentrum, Stefan Meier-Str. 31a, D-79104 Freiburg i. Br., FRG

Prof. Paolo Corradini, Università di Napoli, Dipartimento di Chimica, Via Mezzocannone 4, 80134 Napoli, Italy

Prof. Karel Dušek, Institute of Macromolecular Chemistry, Czech Academy of Sciences, 16206 Prague 616, Czech Republic

Prof: Sam Edwards, University of Cambridge, Department of Physics, Cavendish Laboratory, Madingley Road, Cambridge CB3 OHE, UK

Prof. Hiroshi Fujita, 35 Shimotakedono-cho, Shichiku, Kita-ku, Kyoto 603 Japan

Prof. Gottfried Glöckner, Technische Universität Dresden, Sektion Chemie, Mommsenstr. 13, D-01069 Dresden, FRG

Prof. Dr. Hartwig Höcker, Lehrstuhl für Textilchemie und Makromolekulare Chemie, RWTH Aachen, Veltmanplatz 8, D-52062 Aachen, FRG

Prof. Hans-Heinrich Hörhold, Friedrich-Schiller-Universität Jena, Institut für Organische und Makromolekulare Chemie, Lehrstuhl Organische Polymerchemie, Humboldtstr. 10, D-07743 Jena, FRG

Prof. Hans-Henning Kausch, Laboratoire de Polymères, Ecole Polytechnique Fédérale de Lausanne, MX-D, CH-1015 Lausanne, Switzerland

Prof. Joseph P. Kennedy, Institute of Polymer Science, The University of Akron, Akron, Ohio 44 325, USA

Prof. Jack L. Koenig, Department of Macromolecular Science, Case Western Reserve University, School of Engineering, Cleveland, OH 44106, USA

Prof. Anthony Ledwith, Pilkington Brothers plc. R & D Laboratories, Lathom Ormskirk, Lancashire L40 SUF, UK

Prof. J. E. McGrath, Polymer Materials and Interfaces Laboratory, Virginia Polytechnic and State University Blacksburg, Virginia 24061, USA

Prof. Lucien Monnerie, Ecole Superieure de Physique et de Chimie Industrielles, Laboratoire de Physico-Chimie, Structurale et Macromoléculaire 10, rue Vauquelin, 75231 Paris Cedex 05, France

Prof. Seizo Okamura, No. 24, Minamigoshi-Machi Okazaki, Sakyo-Ku, Kyoto 606, Japan

Prof. Charles G. Overberger, Department of Chemistry, The University of Michigan, Ann Arbor, Michigan 48109, USA

Prof. Helmut Ringsdorf, Institut für Organische Chemie, Johannes-Gutenberg-Universität, J.-J.-Becher Weg 18-20, D-55128 Mainz, FRG

Prof. Takeo Saegusa, KRI International, Inc. Kyoto Research Park 17, Chudoji Minamima-chi, Shimogyo-ku Kyoto 600 Japan

Prof. J. C. Salamone, University of Lowell, Department of Chemistry, College of Pure and Applied Science, One University Avenue, Lowell, MA 01854, USA

Prof. John L. Schrag, University of Wisconsin, Department of Chemistry, 1101 University Avenue. Madison, Wisconsin 53706, USA

Prof. G. Wegner, Max-Planck-Institut für Polymerforschung, Ackermannweg 10, Postfach 3148, D-55128 Mainz, FRG

Table of Contents

Concentrated Emulsion Polymerization
E. Ruckenstein   1

**N-Benzyl and N-Alkoxy Pyridinium Salts as Thermal
and Photochemical Initiators for Cationic Polymerization**
Y. Yagci, T. Endo 59

**Synthesis of Block Copolymers by Radical Polymerization
and Telomerization**
B. Améduri, B. Boutevin, Ph. Gramain 87

Metallocenes for Polymer Catalysis
W. Kaminsky, M. Arndt 143

Author Index Volumes 101–127 189

Subject Index 197

Erratum

Advances in Polymer Science, Vol. 126, contr. Tsuruta, pp. 1–53

Prof. R. Reisfeld is named as the editor of this contribution. However, the correct editor is Prof. H. Ringsdorf.

Concentrated Emulsion Polymerization

Eli Ruckenstein
Chemical Engineering Department, State University of New York
at Buffalo, Buffalo, NY 14260/USA

The present review summarizes results obtained in the last few years in this laboratory regarding the concentrated emulsion polymerization method. In this method, concentrated emulsions have been used as precursors for latexes of homopolymers, copolymers and tough polymers. They have also been employed to prepare conductive polymers, composites, composite membranes, microsponge molecular reservoirs and polymer supported quaternary onium salts, polymer supported palladium complexes and quaternary onium salts, polymer supported enzymes or cells. In contrast to the conventional emulsions, concentrated emulsions have a large volume fraction of dispersed phase, greater than 0.74 and as large as 0.99. When the volume fraction of the dispersed phase is sufficiently large, polyhedral cells of the dispersed phase are separated by thin films of continuous phase. Latexes have been prepared by dispersing a hydrophobic monomer(s) in a small amount of water containing a surfactant, or by dispersing a hydrophilic monomer(s) in a small amount of hydrocarbon (decane) containing a surfactant, and polymerizing the system. Composites have been prepared by dispersing an aqueous solution of a hydrophilic monomer(s) in a small amount of a solution of a hydrophobic monomer in a hydrocarbon, or by dispersing a solution of a hydrophobic monomer(s) in a hydrocarbon in a small amount of a solution of a hydrophilic monomer in water; this was followed by polymerization. Conventional emulsions, microemulsions or colloidal dispersions have been sometimes employed. It is important to emphasize that the polymerized system almost maintains the structure of the emulsion precursor employed.

List of Abbreviations and Tradenames	3
1 Introduction	4
2 Phase Behavior and Stability of Concentrated Emulsions of Hydrocarbons in Water	5
3 Stability of Concentrated Emulsions Containing Monomers	9
4 Polymerization of a Hydrophobic Monomer	18
5 Polymerization of a Hydrophilic Monomer	23
6 Copolymerization of Styrene and Methacrylic Acid in Concentrated Emulsions	26
7 Hydrophilic-Hydrophobic Polymer Composites	29
8 An Improved Concentrated Emulsion Polymerization Pathway . . .	31
9 A Two-Step Colloidal Pathway to Polymer Composites	37

10	Concentrated Emulsion Pathway to Toughened Polymeric Latexes	41
	10.1 Rubber Toughened Polystyrene Composites	42
	10.2 Preparation of MBSB Composites	44
	10.3 Poly(vinylidene chloride)/Poly(butyl methacrylate) Composites Prepared via the Concentrated Emulsion Pathway	45
11	Encapsulation of Solid Particles by the Concentrated Emulsion Polymerization Method	49
12	Concentrated Emulsion Polymerization Pathway to Hydrophobic and Hydrophilic Microsponge Molecular Reservoirs	50
13	Other Applications of Concentrated Emulsions	55
	13.1 Selective Composite Membranes	55
	13.2 Conductive Composite Polymers	55
	13.3 Polymer Supported Catalytic Groups	55
	13.4 Polymer Substrates for the Immobilization of Enzymes and Cells	56
14	References	56

List of Abbreviations and Tradenames

AA	acrylic acid
AAM	acrylamide
Arkopal-N15	α-(4-nonylphenil)-ω-hydroxypoly(oxy-1,2-ethanediyl)
AIBN	azobisisobutyronitrile
BMA	butylmethacrylate
CA	cetyl alcohol
CMC	critical micelle concentration
DSC	differential scanning calorimetry
DMF	dimethyl formamide
DVB	divinyl benzene
EDS	energy dispersive spectroscopy
EMA	ethyl methacrylate
EO	ethylene oxide
HLB	hydrophilic-lipophilic balance
MAA	methacrylic acid
MEHQ	methyl hydroquinone
MMA	methyl methacrylate
OA	oleyl alcohol
PAA	poly(acrylic acid)
PBMA	poly(butyl methacrylate)
PDVB	poly(divinylbenzene)
PMAA	poly(methacrylic acid)
PMMA	poly(methyl methacrylate)
PS	polystyrene
PVBC	poly(vinylbenzene chloride)
PVDC	poly(vinylidene chloride)
SBS	styrene-butadiene-styrene
SDS	sodium dodecyl sulfate
Span 20	sorbitan monolaurate
Span 80	sorbitan monooleate
ST	styrene
THF	tetrahydrofuran
Triton X-100	α-[4-(1,1,3,3-tetramethylbutyl)phenyl]-ω-hydroxypoly(oxy-1,2-ethanediyl)
Tween 20	polyoxyethylene sorbitan monolaurate
VBC	vinylbenzene chloride
VDC	vinylidene chloride

1 Introduction

The goal of this review is to summarize the results obtained in our laboratory regarding the "concentrated emulsion polymerization pathway." This method was used to prepare polymers, copolymers, composites, tough polymers, as well as polymer supported catalysts.

A concentrated emulsion has a large volume fraction of the dispersed phase, larger than 0.74 (which represents the volume fraction of the most compact arrangement of monosized spheres) and as large as 0.99 [1–8]. It has a paste-like appearance and behavior. When the volume fraction of the continuous phase is sufficiently small, the dispersed phase is composed of polyhedral cells separated by thin films of continuous phase (Fig. 1). The concentrated emulsions are prepared by dropwise addition of the dispersed phase to a small amount of a continuous phase containing a surfactant. As is well known, experimental observations have indicated that the phase in which the surfactant is soluble constitutes the continuous phase of an emulsion. The adsorption of the surfactant on the interface between the two media of the emulsion is responsible for its kinetic stability. The concentrated emulsions have attracted our attention as a possible pathway for polymerization for a number of reasons. (i) The ordered organization of the surfactant molecules at the interface between the two media may organize the monomers in its vicinity and thus accelerate the rate of conversion. (ii) The reduced mobility inside the cells, because of the presence of the surfactant layer, could generate an earlier gel effect, which leads to the delay in the bimolecular termination reaction and hence to higher molecular weights. In other words, the rate of polymerization and the molecular weight are expected to be higher if the polymerization occurs in a concentrated emulsion than in bulk. (iii) The polymerized system maintains the structure of the emulsion precursor. (iv) A better control of the size of the latexes can be made by

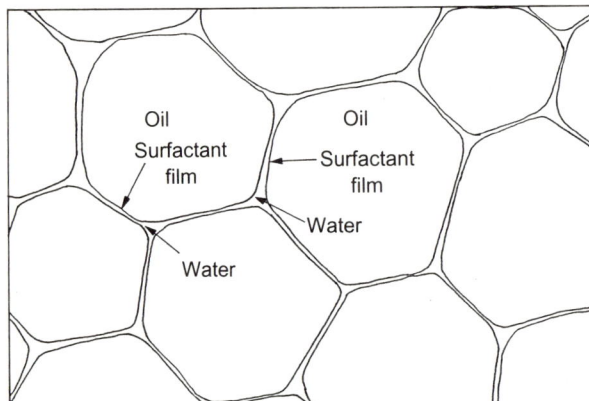

Fig. 1. Sketch of a concentrated emulsion

controlling the size of the cells of the concentrated emulsion, using suitable surfactants, ionic strengths, and pHs. Hence the splitting of a bulk system into a large number of small, independent cells may have favorable effects. In addition, a concentrated emulsion can be polymerized to generate either latexes, or a porous medium or a composite. Because the concentrated emulsions play a major role in the present method, the first two sections are concerned with the concentrated emulsions and their stability. This will be followed by their use in the preparation of polymers, copolymers and composites. Since the stability of the concentrated emulsion precursor is of major importance, some improvements which increase its stability will be examined next. Finally, some applications to tough polymers, encapsulation and to microsponge molecular reservoirs will be presented. A second review will be concerned with the use of concentrated emulsions to generate various kinds of catalysts.

2 Phase Behavior and Stability of Concentrated Emulsions of Hydrocarbons in Water

Before examining the behavior of concentrated emulsions involving polymerizable monomers, it is instructive to provide some general information regarding their phase behavior when hydrocarbons are used as the hydrophobic phase. A small amount of water containing the desired surfactant was located in a test tube in which small amounts of the desired hydrocarbon were added successively with magnetic stirring. The scope of this section is to emphasize that, for given amounts of continuous phase (water) and surfactant dissolved in it, there exists a maximum amount of dispersed phase which can be included in the concentrated emulsion [8a]. Any additional dispersed phase separates as a distinct phase. Double distilled water was used as the continuous phase and n-pentane, n-hexane, n-heptane, isooctane, or decane as the dispersed phase.

In Fig. 2, the weight ratio m_1/m_2 of hydrocarbon to water is plotted as a function of the concentration of surfactant in the continuous phase (in weight percent) when sodium dodecyl sulfate (SDS) is employed as emulsifier. The above ratio is calculated for the point at which a small amount of hydrocarbon remains as a distinct phase. The ratio m_1/m_2 depends upon the nature of the hydrocarbon employed and increases with the surfactant concentration, more rapidly at lower concentrations. A similar behavior was observed for a non-ionic surfactant, Triton X-100 (Fig. 3), but the values of m_1/m_2 are smaller in this case than in Fig. 2. This happens because the electrostatic repulsion responsible for the stability of the concentrated emulsion containing SDS is stronger than the steric repulsion involved in the stability of the emulsion containing Triton X-100.

When the above "saturated" concentrated emulsions were subjected to centrifugation, two phases formed: an optically transparent concentrated emul-

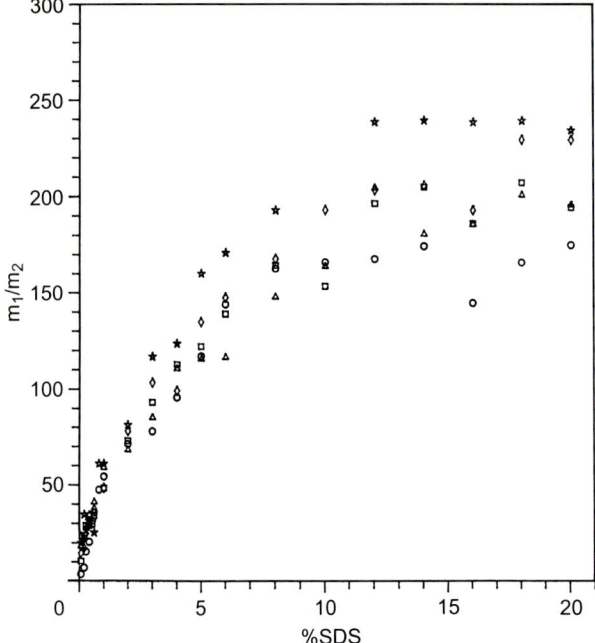

Fig. 2. Plot of m_1/m_2 against weight percentage of surfactant (SDS) in the continuous phase at room temperature: ○, n-pentane; △, n-hexane; □, n-heptane; ◇, isooctane; ∗, decane

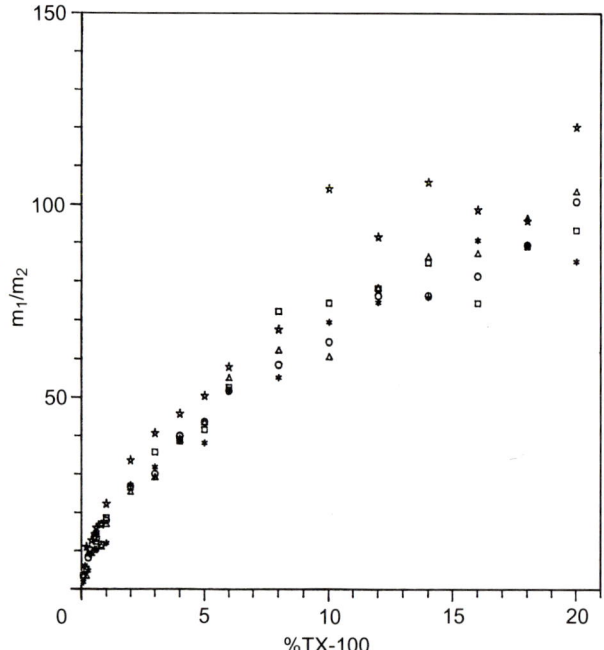

Fig. 3. Plot of m_1/m_2 against weight percentage of surfactant (Triton X-100) in the continuous phase at room temperature: ○, n-pentane; △, n-hexane; □, n-heptane; ∗, cyclohexane; ☆, decane

sion and a small amount of turbid water-rich concentrated emulsion. Strong shearing by shaking broke down the transparent concentrated emulsion into a pure hydrocarbon phase and an emulsion of hydrocarbon in water. In contrast, concentrated emulsions with weight fractions of the continuous phase greater than those corresponding to "saturation" remained stable under shear and regained their structure when shearing was stopped. They flow easily through the needle of a syringe and recover their structure at the exit.

When Arkopal-N15 (Hoechst) was employed as emulsifier and hexane as the dispersed phase (Fig. 4), the ratio m_1/m_2 passed through a maximum for about 16 wt% surfactant and became zero for a concentration of surfactant of about 30%, as a result of the increase in the viscosity of the continuous phase with increasing surfactant concentration. A too high viscosity of the continuous phase probably impedes the incorporation of the hydrocarbon phase into the continuous phase.

The stability of the concentrated emulsions has a kinetic origin. Repulsive double layer forces together with hydration forces are responsible for stability when the surfactant which is adsorbed upon the surface of the thin films is ionic; steric repulsion as well as hydration forces are involved in stability when the adsorbed surfactant is non-ionic.

When SDS was employed as dispersant (Fig. 5), the concentrated emulsion remained stable even in a range of electrolyte concentrations in which the

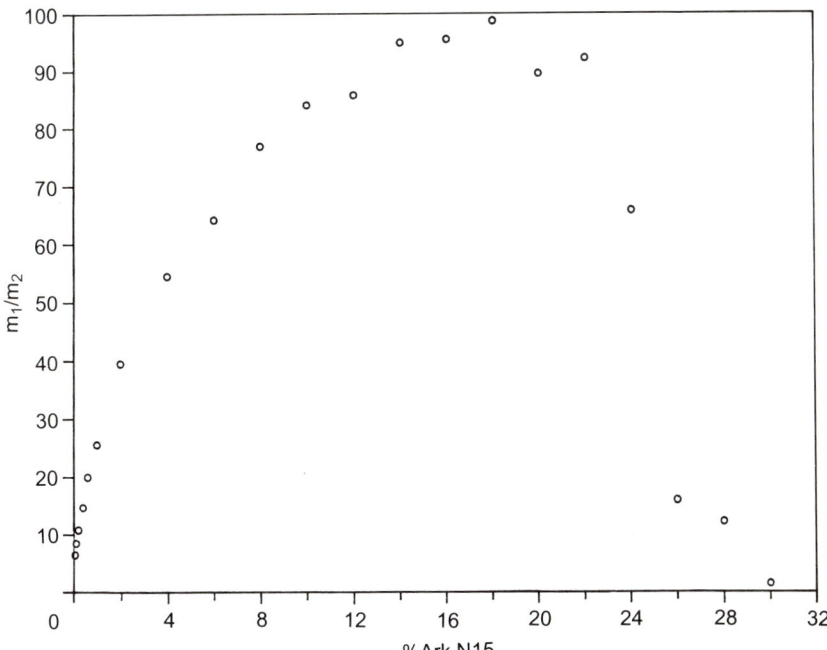

Fig. 4. Plot of m_1/m_2 against weight percentage of Arkopal-N15 in the continuous phase at room temperature. Hexane was the hydrocarbon employed

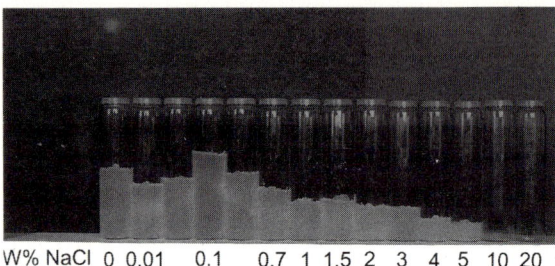

W% NaCl 0 0.01   0.1   0.7 1 1.5 2   3   4   5   10 20
in H₂O   0.05   0.5
                    9.1 W% SDS

Fig. 5. Effect of salt concentration in weight percentage on the amount of gel formed when SDS was employed as emulsifier and hexane as the hydrocarbon

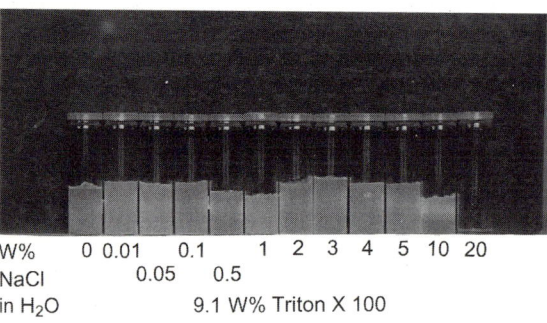

W%   0 0.01   0.1   1   2   3   4   5   10 20
NaCl   0.05   0.5
in H₂O   9.1 W% Triton X 100

Fig. 6. Effect of salt concentration in weight percentage on the amount of gel formed when Triton X-100 was employed as emulsifier and hexane as the hydrocarbon

double layer repulsive forces are expected to be completely shielded. Hence a repulsive force different in nature from double layer forces is acting at sufficiently large salt concentrations [8a]. A possible explanation is as follows. At high ionic concentrations, the free ions and the ion-pairs compete for the water molecules and one can no longer identify hydrated ions as individual ions, but the water and ions become more collectively organized. The presence of an oil-water interface and of the adsorbed surfactant molecules modifies this structure because the adsorbed ionic surfactant molecules form ion-pairs with the counterions. The dipoles of these ion-pairs represent a source of polarization which propagates inside the continuous phase. When two interfaces approach one another, the polarization boundary layers will increasingly overlap, thus decreasing the local average dipole moments, hence increasing the free energy of the system. Consequently a repulsive force is generated.

Concentrated emulsions no longer form above a critical salt concentration, because at such high salt concentrations water and salt are organized to such an extent that the head group of the surfactant molecule no longer has more favorable interactions with the water molecules than with the oil molecules. For this reason the surfactant molecules are salted out from the water into the oil environment and the amount adsorbed on the water-oil interface becomes small. This also indicates that salt alone cannot stabilize the concentrated emulsion.

For a nonionic surfactant (Fig. 6), steric and polarization repulsions are responsible for stability at relatively low surfactant concentrations. Concentrated emulsions are no longer generated as the salt concentration becomes large enough for salting out to occur [8a].

The phase behavior of concentrated emulsions cannot be predicted on the basis of thermodynamic equilibrium, because their stability has a kinetic and not a thermodynamic origin. A concentrated emulsion is stable because the repulsion among its cells is sufficiently strong to delay the rupture of the thin films of continuous phase. Its pseudo "phase" behavior could however be predicted on the basis of drainage equilibrium, which implies a balance between the opposing effects of gravity and capillary forces. There are three possibilities: no phase separation, separation of a single phase (continuous or dispersed), and separation of both phases. A generalized "phase" diagram was suggested [8b, 8c] which can be used to predict the phase behavior in terms of the relative magnitudes of the gravitational and capillary forces.

3 Stability of Concentrated Emulsions Containing Monomers [9]

If the oil phase is replaced in an oil in water (o/w) emulsion by a hydrophobic monomer, or the water phase in a water in oil (w/o) emulsion by a hydrophilic monomer, an emulsion is obtained that can be employed as a precursor for the preparation of polymer latexes [10, 11]. Similarly, if both phases are replaced, the oil phase by a hydrophobic phase containing a monomer and the water phase by a hydrophilic phase containing a monomer, the generated emulsion could be employed as a precursor in the preparation of polymer composites [12]. However, concentrated emulsions that are generated and stable at room temperature may become unstable at the polymerization temperature. To be suitable for the preparation of polymers and polymer composites, the concentrated emulsion must first form and, subsequently, it must remain stable at the temperature at which polymerization takes place. The scope of the present section is to investigate the factors that influence the formation and stability of concentrated emulsions at the preparation and polymerization temperatures in order to identify the physico-chemical conditions that ensure their stability [9].

Stability of emulsions refers to the resistance to the formation of two separate phases [13, 14]. Coalescence of the droplets is responsible for the phase separation. Ostwald ripening constitutes an additional mechanism by which the large droplets grow in size at the expense of the smaller ones, which decrease in size.

The stability of concentrated emulsions is affected by the chemical natures of the dispersed and continuous phases as well as that of the surfactant, the viscosities of the continuous and dispersed phases, the temperature and the volume fraction of the dispersed phase.

Regarding the chemical natures of the dispersed and continuous phases, experiment indicates that the higher the hydrophobicity of one of the phases and the hydrophilicity of the other phase, the more stable is the concentrated emulsion. This observation can be explained as follows. The interactions between strong hydrophobic and hydrophilic phases are relatively weak; consequently, the interfacial free energy between the two phases free of surfactant is expected to be large. The concentrated emulsion is expected to form more easily when the interfacial free energy between the two phases (free of dispersant) is larger [15, 16]. Indeed, in such cases, there are strong interactions between the hydrocarbon tails of the adsorbed surfactant molecules with the hydrophobic medium and of the head groups of the surfactant molecules with the hydrophilic medium, which increase the stability of the emulsion.

The nature of the surfactant and its concentration is expected to play a role. To achieve a mechanically strong interfacial film, which can ensure the stability of the emulsion, the interfacial film of adsorbed surfactant molecules should be condensed in order to have strong lateral intermolecular interactions. A blend of two surfactants with different areas of head groups rather than an individual surfactant can more easily generate a close-packed and mechanically strong interfacial film.

The viscosity of the continuous phase affects the stability of the concentrated emulsion. The viscosity of the continuous phase can be modified either by adding thickeners or by increasing the surfactant concentration. For instance, the formation of a liquid-crystalline structure in the continuous phase when the surfactant concentration is sufficiently large can increase the stability of the emulsion. However, a too high viscosity of the continuous phase caused by a high surfactant concentration can hinder the formation of a concentrated emulsion because it generates resistance to the dispersion of the dispersed phase.

Too small volume fractions of the continuous phase diminish the stability because the thin films which separate the cells of the concentrated emulsion can rupture more easily.

The effect of the nature of the hydrophobic liquid on the formation and stability of concentrated emulsions (measured by the weight percent of bulk phases separated by heating at 50 °C for various times) in which the other phase in water is presented in Fig. 7. As suggested above, the stability of the concentrated emulsion prepared at room temperature and heated subsequently at 50 °C can be correlated with the interfacial tension between the hydrophobic liquid and water measured at 25 °C in the absence of surfactant. As expected from the considerations presented above, a striking increase in stability is observed with increasing interfacial tension between the hydrophobic liquid and water. An increase in the polarity of the hydrophobic organic liquid produces a fall in stability, a too high polarity hindering the formation of the concentrated emulsion because the hydrophobic phase is no longer hydrophobic enough. The surfactant employed in the preparation of the concentrated emulsion affects both the formation and the stability. For the hydrophobic organic liquids with high polarity, for which the interfacial tension at 25 °C is less than about 35 dyn cm^{-1}, the formation of 25 °C and the stability at 50 °C are strongly dependent upon the nature of the surfactant employed. For instance, the

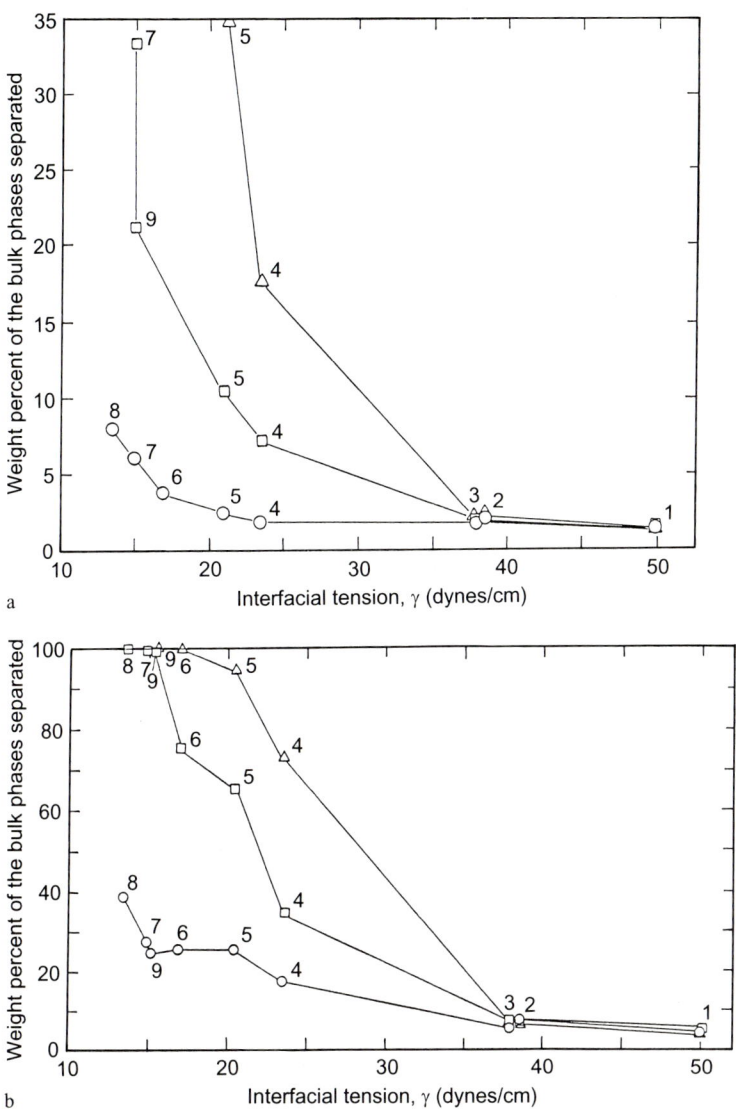

Fig. 7a, b. Weight per cent of bulk phases separated from the concentrated emulsions by heating at 50 °C plotted against the interfacial tension between the hydrophobic liquid and water. The concentrated emulsions were prepared at 25 °C and had a volume fraction of the dispersed phase of 0.9. Weight per cents of bulk phases separated: **a** after 3 h of heating; **b** after 24 h of heating. ○ and □ denote o/w concentrated emulsions prepared using Tween 20 (8.15×10^{-2} mol l^{-1} water) and SDS (3.47×10^{-1} mol l^{-1} water) as surfactant, respectively. △ denotes w/o concentrated emulsions prepared using Span 80 (2.5×10^{-1} mol l^{-1} styrene) as surfactant. Numbers are as follows: 1, decane; 2, ethylbenzene; 3, styrene; 4, *n*-butyl methacrylate; 5, butylacrylate; 6, ethyl methacrylate; 7, vinyl acetate; 8, ethyl acrylate; 9, methyl methacrylate non-ionic surfactant Tween 20 is more effective than the ionic surfactant SDS for o/w concentrated emulsions, which are more stable than the w/o concentrated emulsions. The relatively high polar hydrophobic liquids whose interfacial tensions with water are less than 15 dyn cm^{-1} do not form w/o concentrated emulsions at room temperature when Span 80 is used as surfactant.

In Fig. 8 the interfacial tension between styrene and aqueous acrylic acid solution as well as the stability of the concentrated emulsion is plotted against the concentration of acrylic acid in water. As the concentration of this polar monomer increases, the concentrated emulsion becomes more unstable, and finally the entire concentrated emulsion separates into bulk phases.

The effect of the pH of the aqueous monomer solution on the stability of the concentrated emulsion was examined by partially neutralizing the aqueous solution of acrylic acid with sodium hydroxide. The results are plotted in Fig. 9.

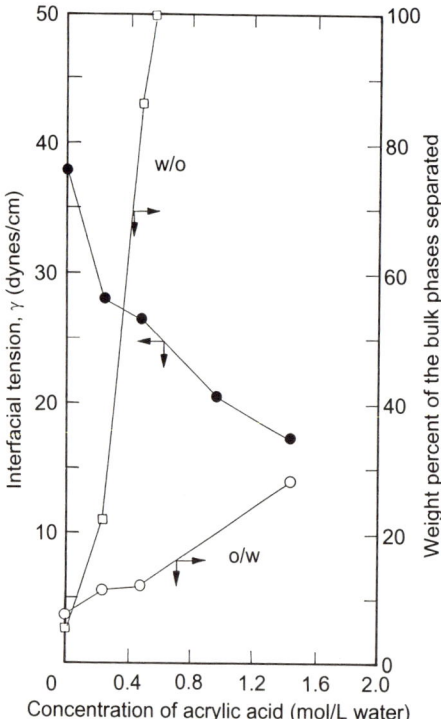

Fig. 8. Interfacial tension between styrene and aqueous acrylic acid solutions and weight per cent of the phases separated from concentrated emulsions by heating for 24 h at 50 °C plotted against the concentration of aqueous acrylic acid solution. The concentrated emulsions were prepared at 25 °C and had a volume fraction of the dispersed phase of 0.9. ● denotes interfacial tension. ○ denotes o/w concentrated emulsions prepared using SDS (3.47×10^{-1} mol l^{-1} water) as surfactant. □ denotes w/o concentrated emulsions prepared using Span 80 (2.5×10^{-1} mol l^{-1} styrene) as surfactant. The w/o concentrated emulsions do not form at 25 °C when Span 20 is used as surfactant and the concentration of acrylic acid is greater than 2.3 mol l^{-1}

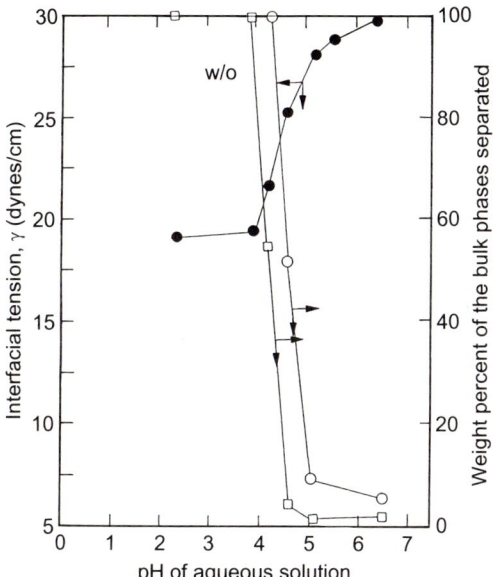

Fig. 9. Interfacial tension between styrene and aqueous acrylic acid solution and weight per cent of bulk phases separated from concentrated emulsions by heating at 50 °C plotted against the pH of the aqueous acrylic acid solution. The concentrated emulsions were prepared at 25 °C and had a volume fraction of the dispersed phase of 0.9. Aqueous acrylic acid solutions (0.94 mol l^{-1} water) were partially neutralized with NaOH. The w/o concentrated emulsions were prepared using Span 80 (2.5 × 10^{-1} mol l^{-1} styrene) as surfactant. ● denotes the interfacial tension. □ and ○ denote weight per cent of bulk phases separated from concentrated emulsions by heating at 50 °C for 3 and 24h, respectively

As the pH of the aqueous solution increases, the interfacial tension and the stability of the concentrated emulsion sharply increase.

In Fig. 10 the interfacial tension and the stability of concentrated emulsions containing styrene and an aqueous sodium chloride solution are plotted against the concentration of sodium chloride. The w/o concentrated emulsions are stable for both Span 20 and Span 80. When SDS was used as surfactant, the o/w concentrated emulsions were more unstable at 50 °C than the above w/o concentrated emulsions because the double layer repulsion between cells is shielded by the high ionic strength. With SDS, concentrated emulsions did not form at room temperature above a salt concentration of 1.2 mol l^{-1} because of the salting-out effect. The o/w concentrated emulsion did not form at all at 25 °C when Span 20 was employed as surfactant.

Table 1 summarizes the relation between the hydrophilic-lipophilic balance (HLB) of surfactants and their ability to form concentrated emulsions. Because the continuous phase is that phase in which the surfactant is soluble, it is expected from the definition of HLB [17, 18] that surfactants with low HLB values are oil-soluble and can therefore generate w/o concentrated emulsions, while those with high HLB values are water-soluble and can lead to o/w concentrated emulsions. Span 20, whose HLB is 8.6, can generate both w/o and o/w concentrated emulsions.

The effect of the HLB of surfactant blends (calculated as the weight average) on the stability of the concentrated emulsions is presented in Fig. 11. (No w/o concentrated emulsion could be prepared using the surfactant blends employed in Fig. 11 because of the phase inversion that occurred at the beginning of the

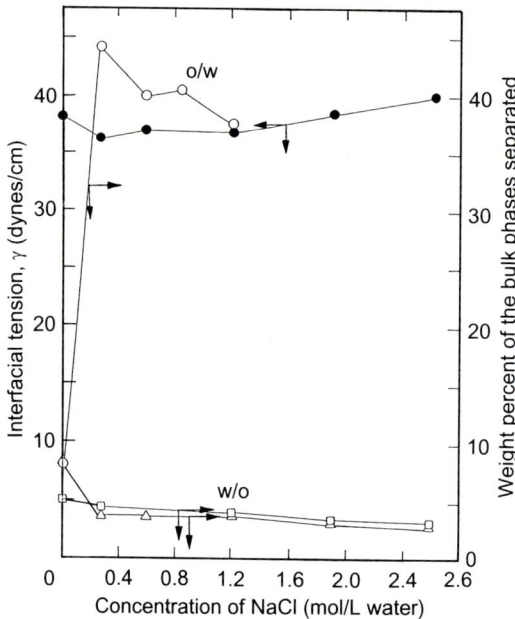

Fig. 10. Interfacial tension between styrene and aqueous NaCl solutions and weight per cent of bulk phases separated from concentrated emulsions by heating for 24 h at 50 °C plotted against the concentration of aqueous NaCl solution. The concentrated emulsions were prepared at 25 °C and had a volume fraction of the dispersed phase of 0.9. ● denotes interfacial tension. ○ denotes o/w concentrated emulsions prepared using SDS $(3.47 \times 10^{-1}\,\text{mol}\,l^{-1}$ water) as surfactant. □ and △ denote w/o concentrated emulsions prepared using Span 80 $(2.5 \times 10^{-1}\,\text{mol}\,l^{-1}$ styrene) and Span 20 $(5.77 \times 10^{-1}\,\text{mol}\,l^{-1}$ styrene) as surfactants, respectively. The o/w concentrated emulsions do not form at room temperature above a salt concentration of $1.2\,\text{mol}\,l^{-1}$ with SDS as surfactant

Table 1. Effect of the HLB of the surfactant on the formation and stability of concentrated emulsions. The concentrated emulsion contains styrene and water as the two phases and the volume fraction of the dispersed phase is 0.9. The concentrated emulsion was prepared at room temperature and its stability test was conducted by heating the emulsion at 50 °C for 3 h and 24 h, respectively

Surfactant[a] (HLB value)	Weight per cent of the bulk phases separated from the emulsion			
	w/o conc. emulsion		o/w conc. emulsion	
	3 h	24 h	3 h	24 h
Span 85 (1.5)	100	100	no emulsion formed	
Span 80 (4.3)	1.8	4.9	no emulsion formed	
Span 40 (6.7)	1.2	4.3	no emulsion formed	
Span 20 (8.6)	1.4	5.6	2.3	10.5
Tween 60 (14.9)	no emulsion formed		1.9	8.0
Tween 40 (15.5)	no emulsion formed		1.7	7.5
Tween 20 (16.6)	no emulsion formed		1.8	6.9
SDS (> 30)	no emulsion formed		2.4	9.1

[a] The concentration of Spans and Tweens were $2.5 \times 10^{-1}\,\text{mol}\,l^{-1}$ and $8.1 \times 10^{-2}\,\text{mol}\,l^{-1}$, respectively. The concentration of SDS was $3.5 \times 10^{-1}\,\text{mol}\,l^{-1}$

emulsification process.) The o/w concentrated emulsions exhibited minimum coalescence at an HLB of about 12. The stability of the concentrated emulsions prepared using surfactant blends is higher than that of those prepared with individual surfactants. When used together, Span and Tween can cover the

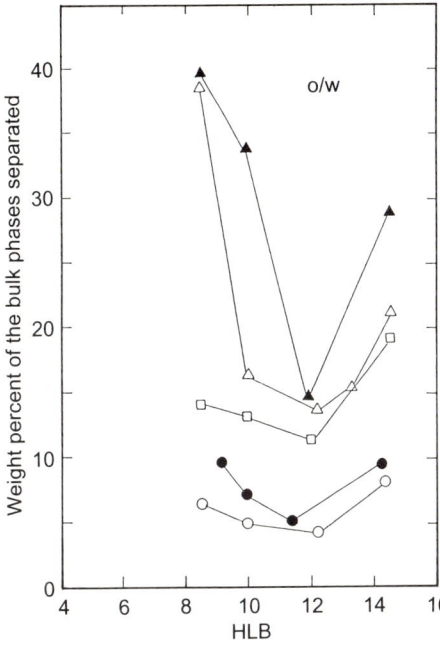

Fig. 11. Weight per cent of bulk phases separated from o/w concentrated emulsions by heating at 50 °C for 24 h plotted against HLB values of surfactant blends. The concentrated emulsions were prepared at 25 °C and had a volume fraction of the dispersed phase of 0.9. The total surfactant blends concentration in each emulsion was held constant at 6.5×10^{-2} mol l^{-1}. ○ and ● denote o/w concentrated emulsions of styrene and water prepared using surfactant blends of Tween 20-Span 85 and Tween 40-Span 85, respectively. □ denotes o/w concentrated emulsions of butyl acrylate and water prepared using surfactant blends of Tween 20-Span 85. △ and ▲ denote o/w concentrated emulsions of methyl methacrylate and water prepared using surfactant blends of Tween 20-Span 20 and Tween 20-Span 85, respectively

interface in a more compact manner because their areas per surfactant molecule are different. Thus the van der Waals forces between the hydrocarbon chains of the surfactants become greater and the interfacial film becomes mechanically stronger.

Table 2 indicates a pronounced effect of the long-chain alcohols on the stability of the concentrated emulsions prepared using a mixture of SDS and long-chain alcohols. The ionic surfactant SDS provides better stabilization when it is coupled with a long-chain alcohol than when it is used alone, because the presence of the alcohol molecules among those of SDS increases the distance between the charged SDS molecules. This decreases the electrostatic repulsion among them and hence increases the cohesion. The cohesion is additionally increased by the more compact packing of the two different species of different sizes. As expected, Fig. 11 and Table 2 also show that the concentrated emulsions prepared using hydrophobic liquids of higher polarity are less stable than those prepared using less polar hydrophobic liquids.

In Fig. 12 the stability of the concentrated emulsions is plotted against the concentration of methyl cellulose in the aqueous continuous phase. Obviously the viscosity of the continuous phase increases with increasing concentration. Concentrated emulsions did not form for high concentrations of methyl cellulose owing to the difficulty of incorporating the dispersed phase into the continuous phase. For instance, styrene-water and n-butyl methacrylate-water did not form concentrated emulsions for methyl cellulose concentrations in the

Table 2. Effect of mixtures of ionic surfactant and long-chain alcohols on the stability of the o/w concentrated emulsions. The concentrated emulsion contains a hydrophobic monomer and water as the two phases. The volume fraction of the dispersed phase is 0.9. The concentrated emulsions were prepared at room temperature and their stability tests were conducted by heating the emulsions at 50 °C for 3 and 24 h, respectively

System	Surfactant[a]/ alcohol	Weight per cent of bulk phases separated from the emulsion	
		3 h	24 h
Styrene/water	SDS	2.4	9.1
	SDS/CA	3.2	8.0
	SDS/OA	1.9	6.3
n-Butyl methacrylate/water	SDS	7.0	44.5
	SDS/CA	3.3	29.1
	SDS/OA	2.7	19.4
Ethyl methacrylate/water	SDS	21.4	100
	SDS/CA	17.6	100
	SDS/OA	13.5	100

[a] When used alone, the concentration of SDS was 3.47×10^{-1} mol l^{-1}; in the mixtures of SDS/CA and SDS/OA, the concentrations of SDS, CA and OA were 1.74×10^{-1} mol l^{-1}

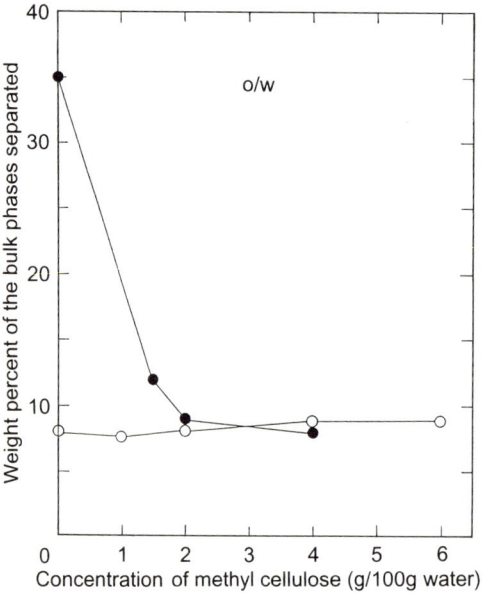

Fig. 12. Weight per cent of bulk phases separated from o/w concentrated emulsions by heating for 24 h at 50 °C plotted against the concentration of methyl cellulose in the aqueous phase containing SDS (3.47×10^{-1} mol l^{-1} water). The concentrated emulsions were prepared at 25 °C and had a volume fraction of the dispersed phase of 0.9. ○ and ● denote styrene and n-butyl methacrylate as the dispersed phases of concentrated emulsions, respectively

aqueous continuous phase higher than 6.2 and 4.5 g/100 g water, respectively. Compared to styrene, the stability of the concentrated emulsions containing the slightly more polar hydrophobic monomer n-butyl methacrylate is increased by the addition of methyl cellulose to the aqueous continuous phase. This might be due to the interactions between methyl cellulose and n-butyl methacrylate.

Figure 13 shows that the stability at 50 °C of concentrated emulsions is affected by the volume fraction of the dispersed phase. As the volume fraction of the dispersed phase increases the stability decreases. The increased instability of the concentrated emulsion is caused by the decreased thickness of the interfacial film that surrounds the cells of the dispersed phase.

In summary, a number of factors are important in the formation at room temperature and stability at the polymerization temperature of 50 °C of concentrated emulsions. They are the chemical nature of the monomers employed in the dispersed and continuous phases, the ability of the surfactants used as dispersants to form strong interfacial films at the interface, the viscosity of the continuous phase, the volume fraction of the dispersed phase, and the temperature. The chemical nature of the hydrophilic and hydrophobic monomers, particularly their polarity, constitutes a major factor. The stability is increased when the interfacial tension between the two phases free of surfactant is higher. The factors that decrease this interfacial tension, such as strong acidity and high concentrations of the hydrophilic monomer in the aqueous phase, have a negative effect on stability. The presence of salt in the continuous phase of the o/w concentrated emulsions destabilizes or even prevents the formation of a concentrated emulsion. Above the CMC, the concentration of the surfactant, if it is not too high, does not affect the stability of the concentrated emulsions. A too high concentration of surfactant in the continuous phase, however, prevents the inclusion of the dispersed phase in the continuous one because it greatly increases the viscosity of the continuous phase. The non-ionic surfactants such as the Tweens can be more effective surfactants than the ionic surfactant SDS. Blends of non-ionic surfactants and mixtures of ionic surfactant and long-chain alcohols are more efficient dispersants than the single surfactants.

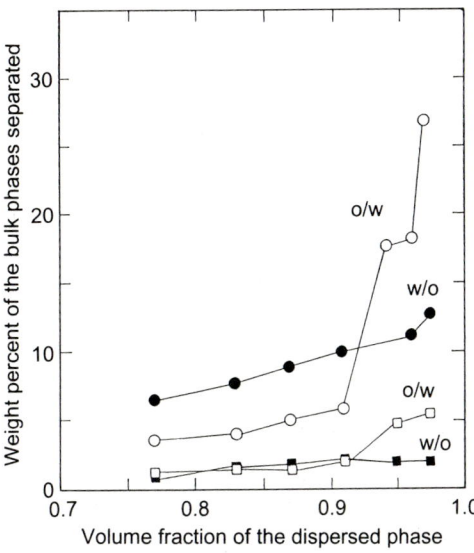

Fig. 13. Weight fraction of bulk phases separated from concentrated emulsions of water and styrene containing either SDS (3.47×10^{-1} mol l^{-1} water) or Span 80 (2.5×10^{-1} mol l^{-1} styrene) by heating at 50 °C plotted against the volume fraction of the dispersed phase. The concentrated emulsions were prepared at 25 °C. □ and ○ denote 3 and 24 h of heating time for the o/w concentrated emulsions, respectively. ■ and ● denote 3 and 24 h of heating time for the w/o concentrated emulsions, respectively

It is almost impossible to obtain a stable concentrated emulsion of a monomer in another one, because their hydrophilicity or hydrophobicity are not high enough. In order to prepare a precursor for a composite, the hydrophobicity of the hydrophobic monomer and/or the hydrophilicity of the hydrophilic monomer must be increased by using their solutions in a hydrocarbon (such as hexane or octane) and water, respectively.

Important information about the structure of concentrated emulsions was gathered by polymerizing either the continuous [2] or the dispersed phase [3].

4 Polymerization of a Hydrophobic Monomer [10]

In the conventional emulsion polymerization, monomer droplets are dispersed in an aqueous phase containing micellar aggregates of surfactant. In this case, the dispersed phase represents a relatively small volume fraction of the system and the micellar aggregates constitute the sites of the polymerization process. In the gel(paste)-like emulsions employed here, the volume fraction of the dispersed phase can be as high as 0.99, and the cells of the concentrated emulsion lead to the polymerized latex particles.

A typical procedure for the preparation of a concentrated emulsion is as follows. A small amount of an aqueous solution containing sodium dodecylsulfate (SDS) was placed in a single neck flask (100 ml capacity) equipped with a mechanical stirrer. Styrene containing the initiator Azobisisobutyronitrile (AIBN) was added to the aqueous solution, with stirring. The whole preparation process of the concentrated emulsion lasted for 10–15 min at room temperature.

The prepared emulsions were transferred to preweighed centrifuge tubes of 15 ml capacity, which were sealed with rubber septa. A mild centrifugation (1500 rpm, less than 1 min) was employed to pack the concentrated emulsions into the tubes when necessary. Polymerizations were conducted in a temperature-controlled water bath in the presence of air. After polymerization, the concentrated emulsion was dispersed by adding water to the tube and agitating with a spatula. The aqueous system was then poured into methanol and the precipitated polymer was separated by filtration.

Figure 14 compares the polymer conversion in concentrated emulsion (gel) polymerization and in bulk polymerization for various polymerization times and for the same concentration of AIBN and temperature, and shows that the conversion is much higher for the concentrated emulsion procedure. The molecular weight (Fig. 15) of the polymer prepared by gel polymerization is higher than that of that prepared by polymerization in bulk by more than one order of magnitude.

Unlike the conventional emulsion polymerization, in which the polymer latexes grow during polymerization, the size and shape of the latexes in concentrated emulsions is principally determined by the preparation of the gel.

Figures 16 and 17 plot log conversion and log M_n (where M_n is the number-average molecular weight) vs log [AIBN] for both polymerizations. In

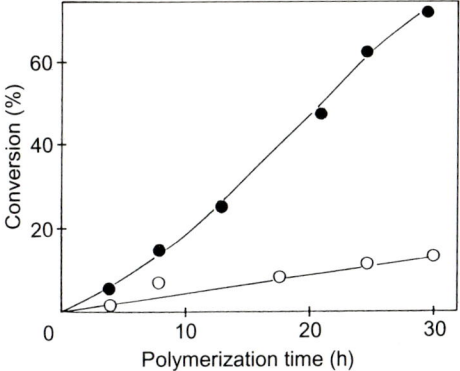

Fig. 14. Plot of conversion against polymerization time (SDS 0.3 g, water 3 ml, styrene 40 ml, 40 °C). (●) Gel; (○) bulk

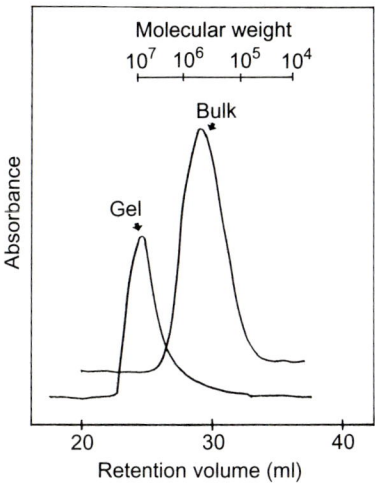

Fig. 15. Molecular weight distribution of polymer in concentrated emulsion polymerization and in bulk polymerization (SDS 0.3 g, water 3 ml, styrene 40 ml, 40 °C, 30 h)

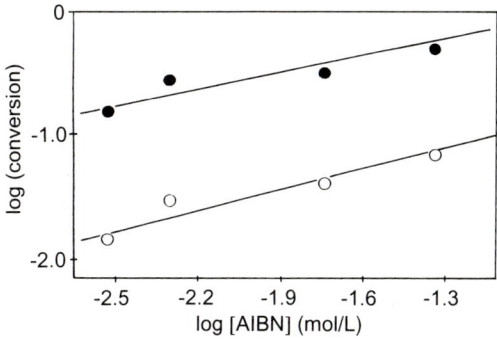

Fig. 16. Plot of log (conversion) against log [AIBN] (SDS 0.3 g, water 2 ml, styrene 15 ml, 40 °C, 10 h). (●) Gel; (○) bulk

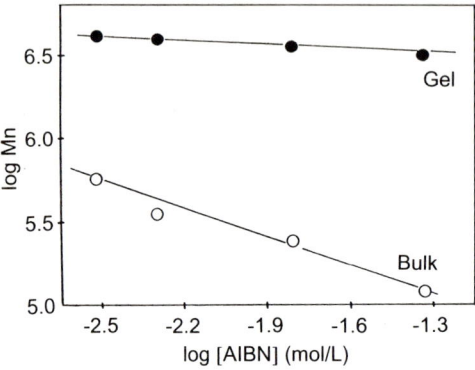

Fig. 17. Plot of log M_n against log [AIBN] (SDS 0.3 g, water 2 ml, styrene 15 ml, 40 °C, 10 h). (●) Gel; (○) bulk

the entire range of initiator concentration, polymerization in concentrated emulsions leads to higher conversions and molecular weights. For bulk polymerization, the slopes for conversion and number-average molecular weight are very close in absolute value, but have opposite signs, results that comply with the kinetics of polymerization. However, in the case of concentrated emulsions, the above relationship between the two slopes is no longer valid. While the rate of polymerization increases with increasing initiator concentration in both cases, the extent of decrease of the molecular weight with increasing initiator concentration is much smaller for the concentrated emulsion polymerization. Because of the emulsifier layer that surrounds the cells, the "mobility" of the liquid within the cells of the gel is lower than in a bulk liquid. As a result, the diffusion of the growing polymer chains is significantly decreased. The termination rate is, therefore, decreased and the rate of polymerization as well as the molecular weight are increased. In other words, the so-called gel effect, which accounts for the abrupt increase in rate and molecular weight of the polymer in the final stages of bulk polymerization because of increased viscosity, comes into play earlier when polymerization is carried out in a concentrated emulsion.

Figure 18, which plots conversion vs monomer volume fraction, exhibits a maximum at about $\phi = 0.9$. This can be explained if one takes into account that, during polymerization, some cells of the gel coalesce and form a bulk phase in which the conversion is smaller. Visual observations indeed indicated a separated thin layer at the upper part of the tube after polymerization. Since no appreciable separated liquid phase was observed before polymerization, it is likely that during polymerization some cells did coalesce. Molecular weight distribution curves have been determined for various values of ϕ. The GPC curves (see Fig. 19) have a tail which is consistent with the molecular weight distribution of the polymer prepared by bulk polymerization. Therefore it is likely that this tail is due to the polymerization in bulk. The greater amount of bulk phase formed for values of ϕ greater than 0.9 is probably due to the decreased stability of the concentrated emulsion in such cases.

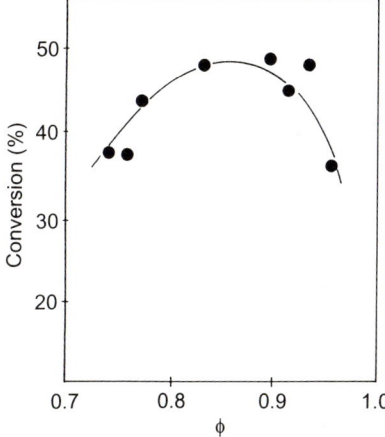

Fig. 18. Plot of conversion against the monomer volume fraction (SDS 0.15 g, styrene 17 ml, 40 °C, 24 h)

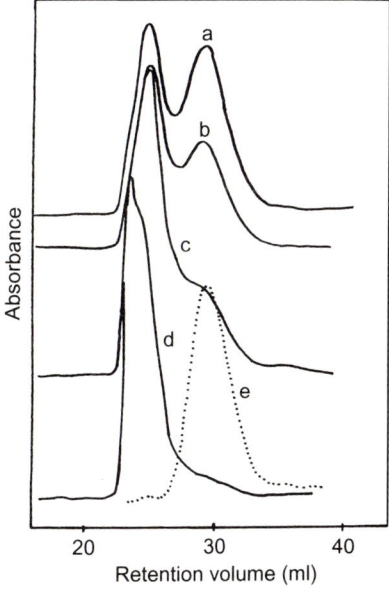

Fig. 19. GPC curves of polystyrene prepared in concentrated emulsions for various SDS concentrations (water 2 ml, styrene 30 ml, 45 °C, 18 h). (a) 1.67×10^{-3} g SDS ml^{-1} styrene; (b) 2.50×10^{-3} g SDS ml^{-1} styrene; (c) 5.00×10^{-3} g SDS ml^{-1} styrene; (d) 1.33×10^{-2} g SDS ml^{-1} styrene; (e) polystyrene obtained by bulk polymerization

Figure 19, which represents GPC curves of the polystyrene prepared in concentrated emulsions of various surfactant concentrations, shows that the amount of bulk phase decreases with increasing surfactant concentration. As the surfactant concentration increases, the weight fraction of polymer formed in the droplets of the gel increases first (Fig. 20) and then levels off, while the conversion shows a linear increase through the entire range of surfactant concentration. Obviously the stability of the gel cells depends on the surfactant concentration. The molecular weight of polymer produced in the gel phase

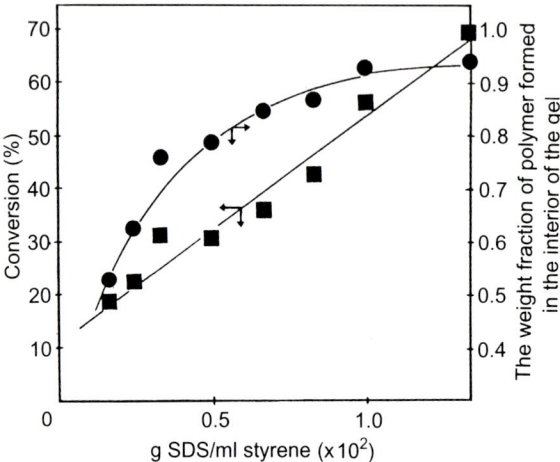

Fig. 20. Plot of conversion and weight fraction of the polymer formed in the interior of the gel as a function of the SDS concentration (water 2 ml, styrene 30 ml, 45 °C, 18 h)

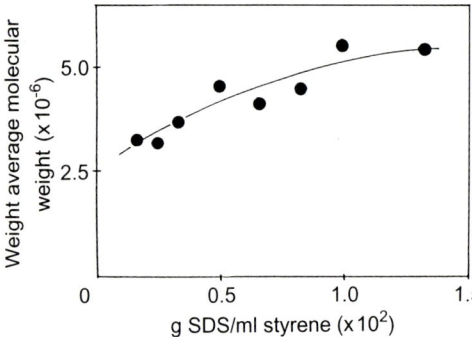

Fig. 21. Plot of the weight average molecular weight of polymer formed in the interior of the gel as a function of the SDS concentration

as a function of the amount of surfactant is plotted in Fig. 21, which shows that the molecular weight of the polymer increases slightly with the amount of surfactant. The electron micrographs (not shown) of the polymer latexes produced with 3.3×10^{-3} g SDS/ml styrene and 1.3×10^{-2} g SDS/ml styrene indicated that the latexes prepared with a larger amount of surfactant contain smaller particles, because a larger interfacial area can be generated. One can speculate that a reduction in the cell size provides higher rigidity in the interior of the droplet, which delays the bimolecular termination step and therefore leads to a larger molecular weight of the polymer and to an increased rate of polymerization. This could account for the continuous increase in conversion with increasing surfactant concentration. It should be mentioned that amounts of surfactant greater than 0.2 g SDS/ml water did not allow the generation of gels, because the high viscosity of the highly concentrated aqueous solutions of surfactant impeded the formation of the cells. Therefore the use of an excessive amount of surfactant should be avoided.

Figure 22 presents the GPC curves of polystyrenes obtained in concentrated emulsions at various temperatures. The molecular weight distribution broadens because of a greater amount of low molecular weight polymers generated in the bulk as the polymerization temperature increases. The greater the temperature, the greater is the coalescence and hence the amount of bulk phase formed.

The bulk phase, reflected in the tail of the GPC curves, increased with NaCl concentration because the double-layer repulsion between the emulsion cells becomes weaker with increasing salt concentration. As a result, the conversion and the weight fraction of polymer formed in the gel phase decrease with increasing NaCl concentration (Fig. 23).

In summary, the polymerization in a concentrated emulsion proceeds faster and produces higher molecular weight polymers than bulk polymerization does. This can be attributed to the gel effect caused by the lower mobility of the species in the emulsion cells.

The minimization of the formation of a bulk phase during polymerization constitutes an important factor in the success of this method, since the rate of the process as well as the molecular weight is much lower in bulk polymerization. A relatively low temperature, a small amount of electrolyte, a suitable amount of water, and an optimum amount of surfactant minimize the occurrence of a bulk phase. The preparation of high molecular weight polystyrene latexes with low size dispersity by concentrated emulsion polymerization was examined in some detail in [19].

5 Polymerization of a Hydrophilic Monomer [11]

The concentrated emulsion polymerization was also applied to an aqueous solution of acrylamide dispersed in decane. Compared to the conventional inverse emulsion polymerization [20], a much smaller amount of organic solvent is employed to produce polymer latexes.

A small amount of decane containing the surfactant Span 80 was placed in a three-neck flask equipped with a mechanical stirrer, an addition funnel, and a nitrogen inlet. The monomer was dissolved in water and placed in the addition funnel. Prior to polymerization, oxygen was removed from the organic and aqueous phases by bubbling with nitrogen for 20 min. A solution containing the initiator was prepared by dissolving $K_2S_2O_8$ in water and by deoxygenating with nitrogen. This solution was injected with a syringe into the aqueous solution of the monomer. The concentrated emulsion was prepared by the dropwise addition of the aqueous solution of monomer to the stirred mixture of decane and surfactant. The final volume fraction of the dispersed phase was 0.94. The preparation of the concentrated emulsion was carried out within 5 min in order to minimize polymerization during the gel preparation step. The polymerization was carried out in a water bath at 40 °C under a nitrogen stream. The

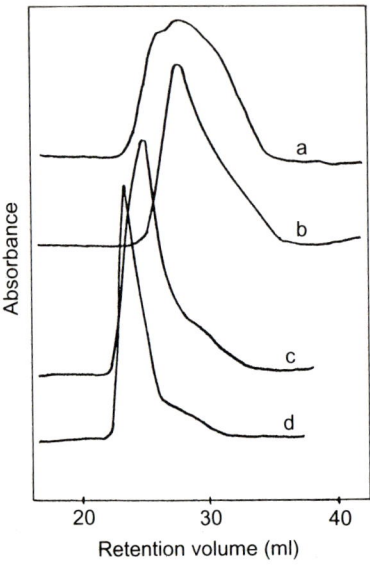

Fig. 22. GPC curves of polystyrene prepared by the concentrated emulsion procedure at various polymerization temperatures (SDS 0.25 g, water 2.5 ml, styrene 42 ml). (a) 65 °C, 20 h; (b) 55 °C, 21 h; (c) 45 °C, 24 h; (d) 35 °C, 72 h

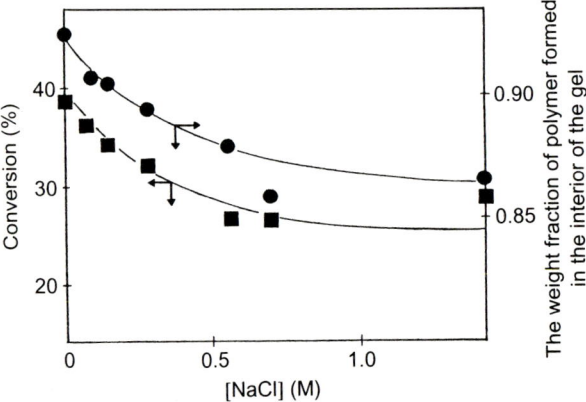

Fig. 23. Plot of conversion and weight fraction of the polymer formed in the interior of the gel as a function of the NaCl concentration in the aqueous phase (SDS 0.225 g, water 1.5 ml, styrene 22.5 ml, 40 °C, 18 h)

polyacrylamide was separated by precipitation in acetone, dried in a vacuum oven, and weighed for the determination of the polymer conversion.

The Span 80 with an HLB (hydrophilic-lipophilic balance number) of 4.3, which is an oil soluble liquid, was used as surfactant. The effect of the continuous medium was investigated by employing 1,1,2,2-tetrachloroethane, toluene, or decane, which have various degrees of hydrophobicity. The amounts of the components used are listed in Table 3. At room temperature (20 °C), concen-

Table 3. The amounts of the components used in the preparation of the concentrated acrylamide emulsions

Component	Amount
Monomer (acrylamide) aqueous solution	40 g
Sorbitan monooleate	1.0 ml
Organic continuous phase[a]	2.5 ml

[a] 1,1,2,2-Tetrachloroethane, toluene or decane

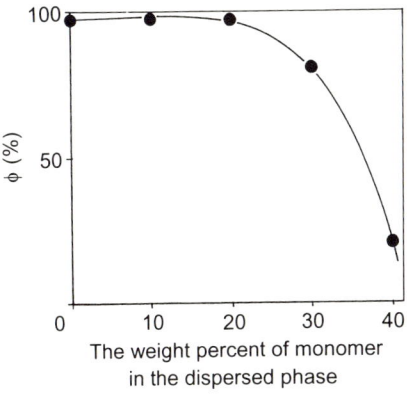

Fig. 24. Plot of weight fraction ϕ of the aqueous solution remaining in the concentrated emulsion state after heating at 60 °C for 5 h against the weight fraction of monomer in the dispersed phase

trated emulsions were formed in toluene and decane for weight fractions of monomer in the aqueous phase as large as 0.4. For 1,1,2,2-tetrachloroethane, concentrated emulsions were obtained only when the weight fraction of monomer was smaller than 0.1. Upon heating at 60 °C for 5 h, the concentrated emulsion prepared at room temperature with toluene separated completely into oil and aqueous layers. The ones based on decane remained, however, largely stable; about 80% of emulsion remained after heating. As expected, a more hydrophobic continuous phase ensures a higher stability of the concentrated emulsion.

The effect of the monomer concentration in the dispersed phase on the stability of the concentrated emulsions, with decane as the continuous medium, prepared at room temperature and heated at 60 °C for 5 h, is presented in Fig. 24. The concentrated emulsion remains stable until a weight fraction of 0.2 is reached; for larger values, the weight fraction which survives as concentrated emulsion decreases and becomes small for a weight fraction of 0.4. The loss of hydrophilicity caused by the increase in the amount of acrylamide accounts for the destabilization of the concentrated emulsion.

The conversion as a function of time is plotted in Fig. 25. The figure also contains results obtained for identical conditions when solution polymerization was employed. Table 4 contains the molecular weights of the polyacrylamide

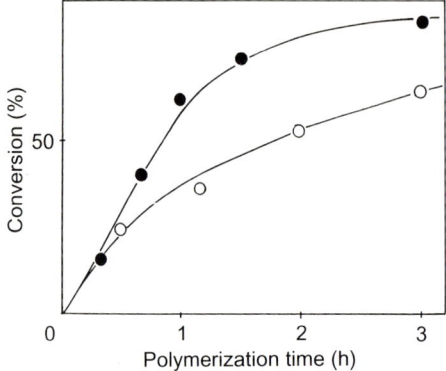

Fig. 25. Conversion as a function of time (in the concentrated emulsion polymerization (●), in the solution polymerization (○)) (acrylamide 12 g, water 28 g, sorbitan monooleate 1.0 ml, decane 2.5 ml, $K_2S_2O_8$ 2.5×10^{-4} g, 40 °C, nitrogen atmosphere)

Table 4. The molecular weight M of polyacrylamide obtained in the concentrated emulsion and in the aqueous solution at various polymerization times

Polymerization time in h	10^{-6} M	
	in solution	in gel
0.5	9.5	7.5
1.5	10.1	7.4

obtained by the concentrated emulsion polymerization and by the solution polymerization in water. The molecular weight is slightly smaller in the concentrated emulsion polymerization. The presence of the surfactant, which has a tertiary alcohol group in its hydrophilic head group, is most likely responsible for the slightly earlier termination by chain transfer of the growing polyacrylamide radical. This is supported by the fact that the chain transfer constant from acrylamide to isopropyl alcohol is quite large (1.9×10^{-3} at 50 °C) [21].

It is important to emphasize that the concentrated emulsion polymerization of acrylamide in water leads to latexes of submicrometer size and employs a very small amount of organic solvent as compared to the conventional emulsion polymerization.

6 Copolymerization of Styrene and Methacrylic Acid in Concentrated Emulsions [22]

In the present section, the copolymerization of styrene and methacrylic acid in a concentrated emulsion in which water containing SDS constitutes the continuous phase is examined. Since the concentrated emulsion requires a small amount of water (typically less than 10% by volume), the loss of methacrylic acid due to its solubility in water can be minimized.

Figure 26 compares the conversion as a function of time in concentrated emulsion and bulk polymerization and shows that polymerization proceeds much faster in a concentrated emulsion. The concentrated emulsion has an internal phase ratio of 0.93 and a molar ratio of MAA/styrene of 0.036. The molecular weight distributions of the polymers generated by both processes are presented in Fig. 27, which shows that concentrated emulsion polymerization leads to molecular weights an order of magnitude higher. Since the copolymer composition changes with conversion, the GPC curves were recorded at the same conversion.

In Fig. 28 the conversion is plotted as a function of the volume fraction of the dispersed phase. This figure shows that as the volume fraction of the dispersed phase decreases, the polymer conversion passes through a maximum at an

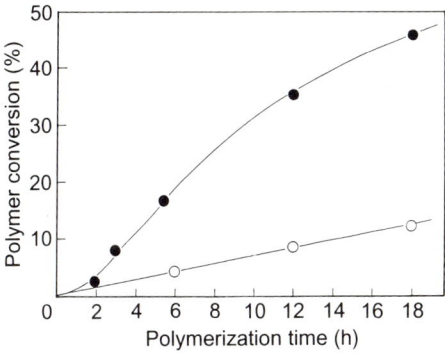

Fig. 26. Conversion as a function of time in a concentrated emulsion (●) and in bulk polymerization (○) (styrene 37 ml, methacrylic acid 1 ml, AIBN 0.3 g, SDS 0.4 g, water 3 ml)

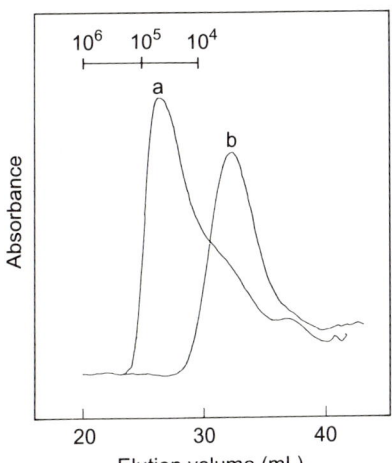

Fig. 27. Molecular weight distribution of the polymers produced in (a) concentrated emulsion (polymerization time = 9 h); (b) bulk polymerization (polymerization time = 41.5 h) (styrene 37 ml, methacrylic acid 1 ml, AIBN 0.3 g, SDS 0.4 g, water 6 ml)

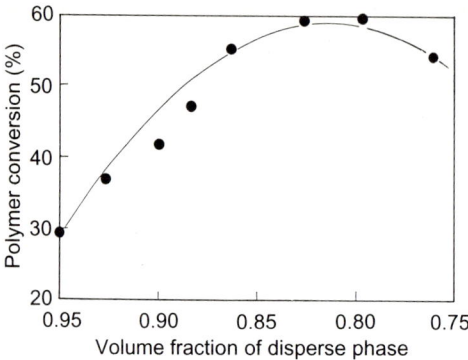

Fig. 28. Conversion as a function of the internal phase ratio (styrene 37 ml, methacrylic acid 1 ml, AIBN 0.3 g, SDS 0.4 g) for a polymerization time of 12h at 40 °C

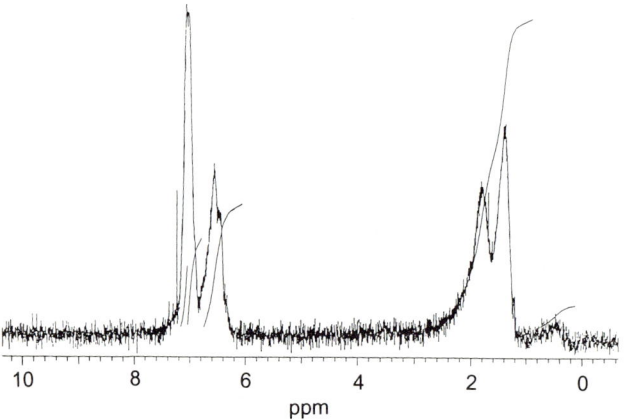

Fig. 29. Proton NMR spectrum of the copolymer latex (in $CDCl_3$) for a mole feed ratio of methacrylic acid to styrene of 0.12, internal phase ratio 0.93, AIBN 0.3 g, SDS 0.4 g and water 3 ml, polymerized first for 12 h at 40 °C. After this polymerization, additional water (twice as much as the weight of the emulsion) was added into the tubes, then polymerization continued for 2 days

internal phase ratio of about 0.82. This happens because some coalescence of the emulsion droplets occurs during polymerization. Polymers are therefore formed both in the gel phase and in the coalesced bulk phase. The increased amount of coalesced phase is responsible, at least in part, for the lower conversion, since polymerization proceeds more slowly in the bulk phase.

Since the methacrylic acid molecules are soluble in both styrene and water, it is important to determine the efficiency of its incorporation in the copolymer latexes. Figure 29 is an H-NMR spectrum of the copolymer latexes. The NMR peak areas of the phenyl (at 6.2–7.4 ppm) and methyl (at about 0.5 ppm) protons allow one to calculate the composition of copolymer latexes. Table 5 lists the compositions and sizes of the copolymer latexes obtained. The efficiency is quite high and increases with increasing amount of methacrylic acid.

Table 5. The composition and size of the copolymer latexes prepared with various molar feed ratios of methacrylic acid (MAA) and styrene[a]

Feed mole ratio of MAA/styrene ($\times 10^2$)	Mole ratio of MAA/styrene in the copolymer latex[b] ($\times 10^2$)	MAA incorporated into copolymer (%)	Average diameter of copolymer latexes[d] (μm)
3.6	3.1	86.1	0.25
7.5	6.7	89.3	0.27
11.5	10.8	93.9	0.27
15.8	14.9[c]	94.3	0.26

[a] The remaining conditions are as in Fig. 29
[b] Calculated on the basis of the NMR peaks
[c] Due to the insolubility in $CDCl_3$, the value was obtained on the basis of the IR spectrum using a calibration curve between IR and NMR
[d] Evaluated from TEM micrographs

7 Hydrophilic-Hydrophobic Polymer Composites [12]

Polymer composites are of practical importance because their two phase structure often allows for a synergistic behavior regarding their mechanical properties and also for selective permeability of organic liquids. A method of preparing a new type of composite by the concentrated emulsion pathway, which will be called hydrophilic-hydrophobic composite, is presented in this section. This composite is synthesized starting from a concentrated emulsion of a hydrophobic (hydrophilic) monomer dispersed in a continuous phase of a hydrophilic (hydrophobic) monomer [12]. In order to increase the hydrophobicity and hydrophilicity of the phases involved, and hence to ensure the stability of the emulsion (gel), solutions of the corresponding monomers in decane (for the hydrophobic monomer) and water (for the hydrophilic monomer) have been employed. The gel was stabilized by introducing a suitable surfactant into the continuous phase. Initiators were added to each of the phases before the preparation of the concentrated emulsion (gel). After the preparation of the gel at room temperature, the system was heated to 40 °C for polymerization to occur. This method can be contrasted to the interpenetrating network method in which a crosslinked polymer is swelled with a second monomer and with crosslinking and activating agents, and finally subjected to a second polymerization.

Two types of composites have been prepared.

I. Polystyrene (Dispersed Phase)-Polyacrylamide (Continuous Phase) Composite (Table 6), and
II. Polyacrylamide (Dispersed Phase)-Polystyrene (Continuous Phase) Composite (Table 7).

Table 6. Amounts of components used in the preparation of the polymer composite

The dispersed phase	
Styrene	27 g
Initiator (AIBN)	2.0×10^{-4} g g^{-1} of styrene
The continuous phase	
Acrylamide	1.0 g
Water	4 ml
Initiator (sodium persulfate)	1.7×10^{-4} g g^{-1} of acrylamide
Surfactant (sodium dodecylsulfate, SDS)	0.3 g

Table 7. The amounts of the components used in the preparation of the polyacrylamide polystyrene composite

Dispersed phase	
12 g acrylamide and 28 g water initiator (sodium persulfate)	1.7×10^{-4} g g^{-1} of acrylamide
Continuous phase	
Styrene	2.5 g
Initiator (AIBN)	2.0×10^{-4} g g^{-1} of styrene
Decane	2 ml
Surfactant (sorbitan monooleate)	1 ml

The effect of the amount of styrene present in the continuous phase on the stability of the gel is presented in Fig. 30. It should be noted that in the stability experiments the initiators were not introduced into the phases. The gel remained stable until the weight fraction of styrene in the continuous phase reached about 60% and the stability decreased very rapidly beyond this value. This happens because styrene is not sufficiently hydrophobic.

The weight percent of bulk phases which separated after heating the styrene (dispersed phase)-acrylamide (continuous phase) gel at 40 °C for 22 h is plotted in Fig. 31 vs the weight percent of acrylamide in water. It increased as the concentration of acrylamide increased; the gel became completely unstable when the amount of acrylamide in the continuous phase reached about 50 wt%. Gels did not form even at room temperature when the amount of acrylamide was greater than this value. The loss of hydrophilicity caused by the increase in the amount of acrylamide accounts for the destabilization of the gel.

A transmission electron micrograph of a composite based on the concentrated emulsion precursor of Table 6 is presented in Fig. 32. Polystyrene

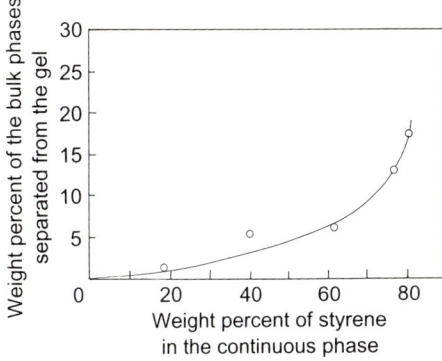

Fig. 30. Weight percent of the bulk phases separated from the gel by heating, at 40 °C for 24 h, the gel prepared at room temperature against the weight percent of styrene in the continuous phase. The composition of the gel is that from Table 7 with the exception of the amount of styrene

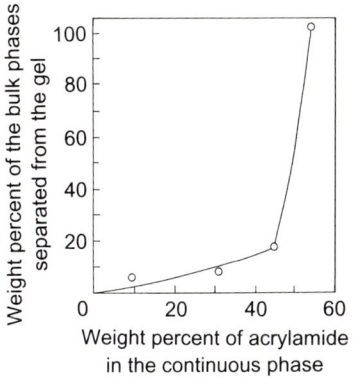

Fig. 31. Weight percent of the bulk phases separated from the gel by heating, at 40 °C for 22 h, the gel prepared at room temperature against the weight percent of acrylamide in the continuous phase. The composition of the gel is that from Table 6 with the exception of the amount of acrylamide

particles can be detected which are nearly spherical and are separated by polyacrylamide films. The material obtained has the appearance of a light blue solid.

A transmission electron micrograph of a composite whose composition is given in Table 7 is presented in Fig. 33. This figure shows that the dispersed phase is composed of polyhedral cells of polyacrylamide separated by films of polystyrene. The obtained material is white, soft, and exhibits some elasticity.

8 An Improved Concentrated Emulsion Polymerization Pathway [23]

The main condition for the use of concentrated emulsions as precursors for polymers, copolymers, and polymer composites is the stability of the gel at the

Fig. 32. Transmission electron micrograph of a composite with the composition given in Table 6

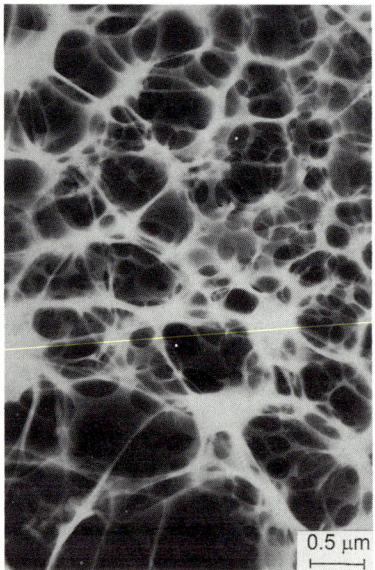

Fig. 33. Transmission electron micrograph of a composite with the composition given in Table 7. *The white regions* represent the polyacrylamide and *the black regions* the polystyrene

polymerization temperature. So far only a few monomers such as styrene, divinylbenzene, and acrylamide could be employed to prepare gels stable both at the preparation (20 °C) and polymerization (50 °C) temperatures. An improvement can, however, be achieved by using a two-step method. In the first step, the monomer is partially polymerized in bulk by heating at 50 °C until a suitable (relatively small) conversion is reached. Furthermore, the partially polymerized monomer is used as the dispersed phase of a concentrated emulsion. We found that monomers that could not lead to stable concentrated emulsions could generate such gels after partial polymerization.

In the first step, the hydrophobic monomer was partially polymerized in bulk, using AIBN as initiator, by heating at 50 °C until a small conversion was reached. Then the inhibitor methylhydroquinone (MEHQ) was introduced to terminate polymerization. In the second step, the partially polymerized monomer was injected at room temperature into a small amount of a stirred aqueous solution of SDS to generate a concentrated emulsion.

The first step in the preparation of a polymer was the same as that used for the preparation of a two-step concentrated emulsion, but without introducing MEHQ at the end of the partial polymerization step. Subsequently the concentrated emulsion was subjected to polymerization by heating at 50 °C for 50 h.

The amounts of the components involved in the concentrated emulsions are listed in Table 8. The weight percent of bulk phases separated from the concentrated emulsions prepared at room temperature and heated at 50 °C for 24 h is considered a measure of its stability. The weight percent of bulk phase separated from the concentrated emulsions prepared in one step at room temperature, in the absence of initiator, and subsequently heated for 24 h at 50 °C is presented in Table 9, which also contains the values of the interfacial tension, γ, measured at room temperature, between the hydrophobic monomer and water, in the absence of surfactant. It should be noted that the stability of the concentrated emulsion is almost unaffected by the presence of AIBN and MEHQ. Table 9 shows that in the one-step method the stability is highest for styrene and lowest for MMA, and that the stability increases with increasing interfacial tension between water and monomer in the absence of surfactant.

Figures 34–36 show that the concentrated emulsions prepared by a two-step procedure are much more stable – the longer the initial polymerization time

Table 8. Amounts used in the preparation of the concentrated emulsion

Dispersed phase: hydrophobic monomer[a] (MMA, EMA, BMA, EHA, ST)	13.5 ml
Continuous phase: water	1.5 ml
Surfactant: SDS	0.152 g
Initiator[b]: AIBN	0.033 g
Inhibitor[c]: MEHQ	0.010 g

[a] Methyl methacrylate (MMA), ethyl methacrylate (EMA), butyl methacrylate (BMA), ($\pm$) 2-ethyl hexylacrylate (EHA) and styrene (ST)
[b] AIBN was used in the partial polymerization of the monomer
[c] MEHQ was used to terminate the polymerization in the partial polymerization

Table 9. Stability of the concentrated emulsion prepared in a one-step method and the interfacial tension between monomer and water in the absence of surfactant

Monomer	MMA	EMA	BMA	EHA	ST
ϕ (%)	100[a]	80	40	14	8
γ (dyn cm^{-1})	16.38	17.69	23.90	24.74	37.41

[a] The concentrated emulsion of MMA in water fully separated in 3 h by heating at 50 °C

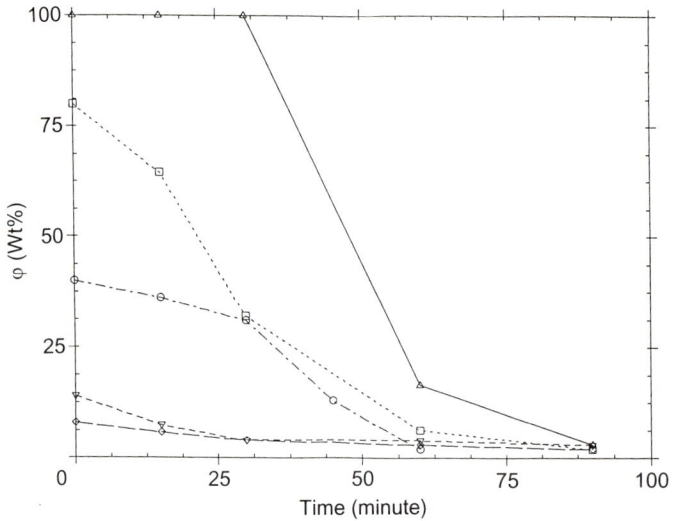

Fig. 34. Weight percent ϕ of bulk phases separated from concentrated emulsion after heating for 24 h at 50 °C against partial polymerization time of monomer: ($\Diamond$) ST, ($\triangle$) MMA, ($\square$) EMA, ($\bigcirc$) BMA, ($\triangledown$) EHA

during the first step, the more stable the concentrated emulsion. The stability of the concentrated emulsion depends on the viscosity of the dispersed phase. When a sufficiently high viscosity is reached, even MMA can generate stable concentrated emulsions. However, the conversion and hence the viscosity should not be too large, since in such cases it is difficult to disperse one of the phases in the other.

The concentrated emulsion free of MEHQ prepared at room temperature can be further polymerized by heating at 50 °C for 50 h. Several polymers, copolymers, and polymer composites were prepared by the two-step method. The amounts of components employed to prepare copolymers and polymer composites are listed in Table 10, whereas the amounts of components employed to prepare polymers are those in Table 8 (of course without MEHQ). The one-step pathway of the corresponding materials was also employed. In the latter case, very high phase separations occurred in most systems, except those

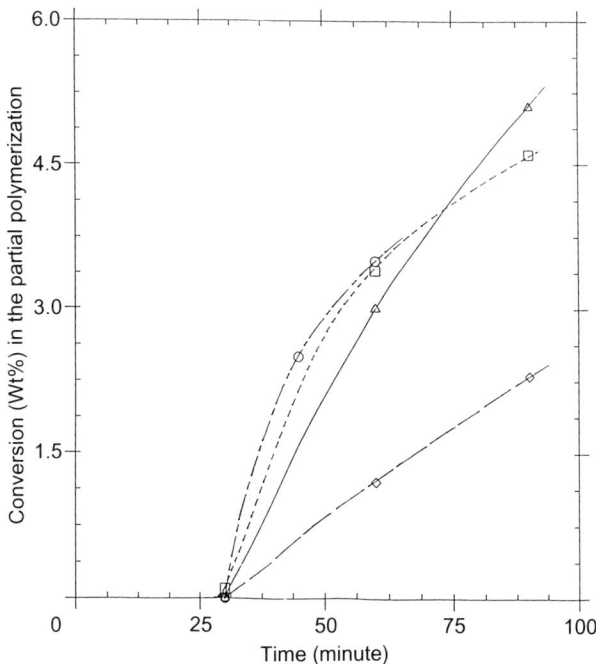

Fig. 35. Polymer conversion (wt%) against partial polymerization time of monomer: (◇) ST, (△) MMA, (□) EMA, (○) BMA

employed to prepare polystyrene (PS) and the PS/PAAM composite. PMMA, PEMA, and the PBMA/PAAM composite could not be prepared by the one-step concentrated emulsion polymerization. Figures 35 and 36 show that the conversion of the monomer and the viscosity do not change in a detectable way during the first 30 min, whereas in Fig. 34 the stability increases gradually in this time interval. The adsorption of small amounts of the produced polymer on the interface between the dispersed and continuous media probably contributes, via steric stabilization, to increased stability.

Two factors contribute to the stability of the gels prepared by the two-step concentrated emulsion. The repulsive forces between the charged surfactant molecules adsorbed on the surface of neighboring cells of the dispersed phase is one of them. The increased viscosity of the dispersed phase which contains the monomer constitutes the second factor, since the increased viscosity opposes the separation of the phases. The partial polymerization increases the viscosity of the dispersed phase, thus increasing the stability of the concentrated emulsion. Monomers that could not lead to gels in the one-step concentrated emulsion method were able to generate them when the two-step pathway was employed. Using this pathway, almost all monomers could be employed to prepare polymer materials.

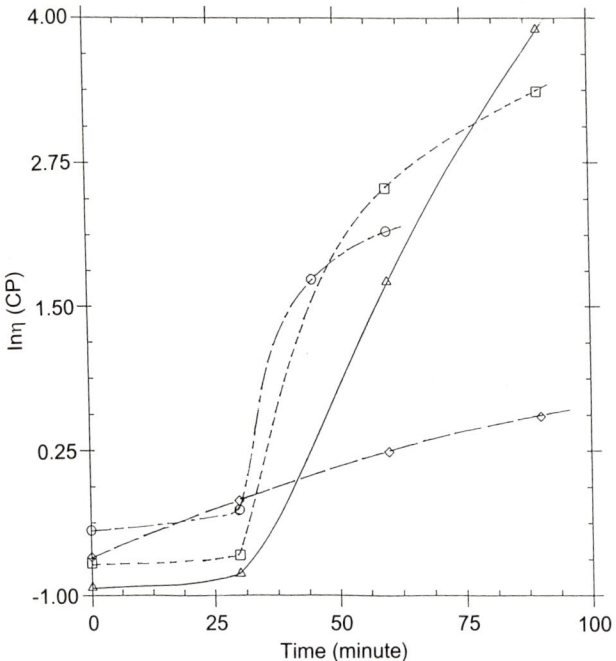

Fig. 36. Logarithm of the viscosity of the partially polymerized monomer against partial polymerization time: (◇) ST, (△) MMA, (□) EMA, (○) BMA

Table 10. Amounts of components used in the preparation of copolymers and polymer composites

Hydrophobic-Hydrophobic Copolymer[a]	
Dispersed phase: ST	6.75 ml
other monomer (MMA, EMA, BMA)	6.75 ml
Continuous phase: water	1.5 ml
Initiator in the dispersed phase: AIBN	1.5×10^{-2} mol l^{-1} monomer
Surfactant: SDS	0.152 g
Hydrophobic-Hydrophilic Polymer Composite[b]	
Dispersed phase: hydrophobic monomer	
(ST or BMA)	13.5 ml
Initiator in the dispersed phase: AIBN	1.5×10^{-2} mol l^{-1} monomer
Continuous phase: water	1.5 ml
and AAM	0.24 mol l^{-1} water
Initiator in the continuous phase:	
$K_2S_2O_8$	4.02×10^{-4} g ml^{-1} water
Surfactant: SDS	0.152 g

[a] The copolymers were prepared by the two-step polymerization; MMA, EMA, and BMA were partially polymerized for 1.5 h
[b] The polymer composites were prepared by the two-step polymerization. ST and BMA were partially polymerized for 1.0 h

9 A Two-Step Colloidal Pathway to Polymer Composites [24]

Polymer composites (blends) are prepared to control the mechanical properties of the polymeric materials and for obtaining permselective membranes for separation processes. Many polymer composites are, however, blends of incompatible polymers. Because of the tendency for segregation caused by the incompatibility, these materials can acquire a nonuniform structure. It is, therefore, useful to develop blends whose structure can be more uniform.

In the methodology developed by us [24], the incompatibility of the two polymers was exploited in a positive way. The composites were obtained using a two-step method. In the first step, hydrophilic (hydrophobic) polymer latex particles were prepared using the concentrated emulsion method. The monomer-precursor of the continuous phase of the composite or water, when that monomer was hydrophilic, was selected as the continuous phase of the emulsion. In the second step, the emulsion whose dispersed phase was polymerized was dispersed in the continuous-phase monomer of the composite or its solution in water when the monomer was hydrophilic, after a suitable initiator was introduced in the continuous phase. The submicrometer size hydrophilic (hydrophobic) latexes were thus dispersed in the hydrophobic (hydrophilic) continuous phase without the addition of a dispersant. The experimental observations indicated that the above colloidal dispersions remained stable. The stability is due to both the dispersant introduced in the first step and the presence of the films of the continuous phase of the concentrated emulsion around the latex particles. These films consist of either the monomer-precursor of the continuous phase of the composite or water when the monomer-precursor is hydrophilic. This ensured the compatibility of the particles with the continuous phase. The preparation of poly(styrenesulfonic acid) salt latexes dispersed in cross-linked polystyrene matrices as well as of polystyrene latexes dispersed in crosslinked polyacrylamide matrices is described below. The two-step method is compared to the single-step ones based on concentrated emulsions or microemulsions.

In the preparation of poly(styrenesulfonic acid) salt latex particles, the concentrated emulsions prepared at room temperature were relatively stable, since their heating at 40 and 50 °C generated only small amounts of bulk phases. The amounts of various components employed in the first polymerization step are listed in Table 11 and the conversion as a function of time is plotted in Fig. 37. Also included in this figure is the conversion, under the same conditions, when solution polymerization was employed. Table 12 shows that the molecular weights of the polymers obtained by concentrated emulsion polymerization are slightly larger than those obtained by solution polymerization. The scanning electron micrographs showed that spherical latex particles of submicrometer sizes in the 0.1–1.0 µm size range were obtained.

For the preparation of polystyrene latexes, the compositions listed in Table 11 were employed. The molecular weight and conversion after 24 h of

Table 11. Representative compositions in the preparation of poly(styrenesulfonic acid) salt and poystyrene latexes

Composition for the preparation of poly(styrenesulfonic acid) salt latexes

Dispersed phase
    styrenesulfonic acid sodium salt   9 g
    water   40 g
    initiator (sodium persulfate)   0.008 g g^{-1} of styrenesulfonic acid salt

Continuous phase
    styrene   3 ml
    initiator (AIBN) 2.0×10^{-2} mol l^{-1} of styrene
    surfactant (sorbitan monooleate)   1.5 ml

Composition for the preparation of polystyrene latexes
Dispersed phase
    styrene   1.5 ml
    initiator (AIBN) 1.26×10^{-2} mol l^{-1} of styrene

Continuous phase
    water   2 ml
    surfactant (SDS)   0.3 g

Table 12. Molecular weight of poly(styrenesulfonic acid) salt prepared at 40 °C by the concentrated emulsion and solution methods

Polymerization time, h	Concentrated emulsion	Mol wt solution
3	0.89×10^6	0.40×10^6

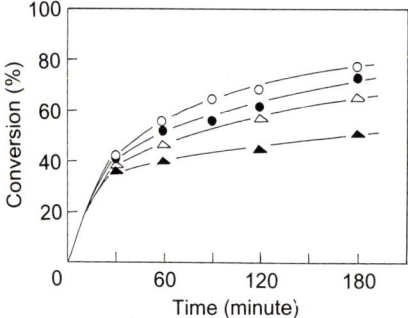

Fig. 37. Polymer conversion as a function of time in the concentrated emulsion at 50°C (●) and 40°C (▲) and in the solution polymerization at 50°C (○) and 40°C (△)

polymerization were 3.1×10^6 and 65%, respectively. The scanning electron micrographs showed that the sizes of the latex particles were about 0.1 µm.

The compositions of the prepared polymer composites are listed in Tables 13 and 14.

Table 13. Representative compositions in the preparation of polymer composites: polystyrene particles dispersed in cross-linked polyacrylamide

H_1	polystyrene latexes[a]	15 g
	acrylamide	1.5 g
	N,N'-methylenebisacrylamide	0.1 g
	initiator (potassium persulfate)	0.01 g g^{-1} of acrylamide
	water	1.5 g
H_2	polystyrene latexes[a]	10 g
	acrylamide	1.5 g
	N,N'-methylenebisacrylamide	0.1 g
	initiator (potassium persulfate)	0.01 g g^{-1} of acrylamide
	water	1.5 g

[a] The compositions of the polymer latexes are given in Table 11

Table 14. Representative compositions in the preparation of polymer composites: poly(styrenesulfonic acid) salt particles dispersed in cross-linked polystyrene

l_1	poly(styrenesulfonic acid) salt latexes[a]	15 g
	styrene	1.6 g
	divinylbenzene	0.4 g
	initiator (AIBN)	0.02 g g^{-1} of styrene
l_2	poly(styrenesulfonic acid) salt latexes[a]	8 g
	styrene	1.6 g
	divinylbenzene	0.4 g
	initiator (AIBN)	0.02 g g^{-1} of styrene

[a] The compositions of polymer latexes are given in Table 11

Figure 38 plots the swelling of the polymer composites composed of polystyrene latexes dispersed in cross-linked polyacrylamide in a mixture of cyclohexane and toluene. The mixture toluene-cyclohexane was chosen for the swelling test because polyacrylamide, which is hydrophilic, is insoluble in both components, and because the benzenic ring makes polystyrene compatible with toluene. The swelling, which in this case is mainly due to the swelling of polystyrene by toluene, is a measure of the strength of the polyacrylamide continuous phase. It is also a measure of the capability of the composite to separate toluene from cyclohexane (aromatics from paraffinics). The compositions of polymer composite films used in the swelling experiments are given in Table 13. The swelling of the film of composition H_1 is about 3.2 in toluene and 1.1 in cyclohexane. As expected, the swelling increases with the fraction of polystyrene latexes in the composite and decreases with increasing cyclohexane concentration. This occurs because of the higher solubility of toluene in polystyrene. The composites I_1 and I_2 (Table 14), in which poly(styrenesulfonic acid) salt latexes are dispersed in cross-linked polystyrene, exhibit selective swelling for water from mixtures of water-ethanol because of the presence of the hy-

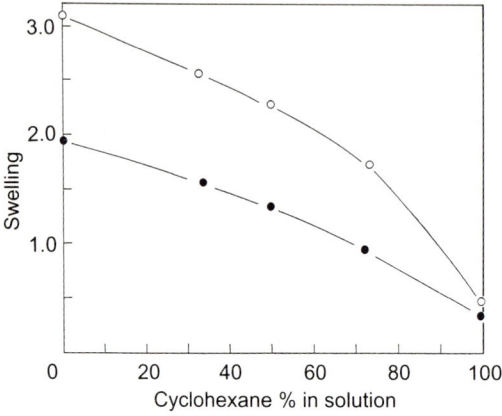

Fig. 38. Swelling as a function of cyclohexane concentration for a cyclohexane–toluene mixture at 25 °C. ○ and ● denote the polymer composites H_1 and H_2 (Table 13), respectively

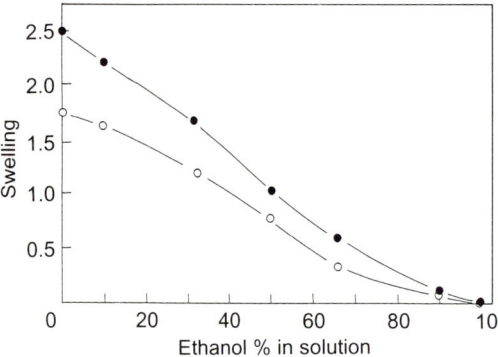

Fig. 39. Swelling as a function of ethanol concentration in an ethanol–water mixture at 25 °C. ○ and ● denote the polymer composites I_1 and I_2 (Table 14), respectively

drophilic poly(styrenesulfonic acid) salt in the composites. The swelling of the composites I_1 and I_2 in ethanol-water mixtures is plotted in Fig. 39. The swelling increases with the fraction of poly(styrenesulfonic acid) salt latexes in the composite and decreases with the concentration of ethanol in the mixture.

Single-step preparations of composite polymers have been examined in previous sections. The volume fraction of the continuous phase was, however, relatively small in those cases. In contrast, the present method allows us to prepare composites with larger volume fractions of the continuous phase. Composites with large volume fractions of the continuous phase can also be obtained in a single-step by polymerizing an emulsion or a microemulsion [24]. An emulsion of a hydrophobic (hydrophilic) monomer in another hydrophilic (hydrophobic) monomer can be extremely stable (even thermodynamically stable, and then it is called a microemulsion) if a sufficiently large amount of surfactant is introduced into the system. For an emulsion to be thermodynamically stable, a cosurfactant is in most cases needed besides the surfactant. The latter method was used to prepare composites by employing acrylamide (hydrophilic) and methyl methacrylate (hydrophobic) as monomers and SDS or aerosol OT as surfactants. The compositions are listed in Table 15. No phase segregation was observed in the composites J_1 and J_2 thus obtained. The composite J_1 is completely transparent like the initial dispersion. The two-step colloidal pathway uses much smaller amounts of low molecular weight surfactants compared to the microemulsion pathway. As a result, the two-step colloidal pathway is expected to lead to composites with better mechanical strength than those obtained by the single emulsion (microemulsion) step.

10 Concentrated Emulsion Pathway to Toughened Polymeric Latexes

Physical or chemical modification methods have been employed to increase the toughness of polymer materials. The chemical modifications include random copolymerization, block copolymerization, grafting, etc.; the physical ones include blending, reinforcing, filling, interpenetrating networks etc. [24–26].

In what follows, a number of methods developed by our group to prepare toughened polymeric composites, which involve both physical and chemical modifications and are based on the concentrated emulsion pathway, are presented [27–29].

Table 15. Representative compositions for composites by one-step microemulsion polymerization

J_1: Compositions of the composites in which polyacrylamide is dispersed in a poly(methyl methacrylate) matrix

Continuous phase	
acrylamide	4 g
water	10 g
initiator (sodium persulfate)	0.01 g g^{-1} of acrylamide
surfactant (SDS)	2 g
Dispersed phase	
methyl methacrylate	3 g
initiator (AIBN)	0.03 g g^{-1} of methyl methacrylate

J_2: Compositions of the composites in which poly(methyl methacrylate) is dispersed in a polyacrylamide matrix

Continuous phase	
methyl methacrylate	10 g
initiator (AIBN)	0.03 g g^{-1} of methyl methacrylate
surfactant (Aerosol OT)	3 g
Dispersed phase	
acrylamide	0.6 g
water	1.4 g
initiator	0.01 g g^{-1} of acrylamide

10.1 Rubber Toughened Polystyrene Composites [27]

A blend was prepared by dissolving a rubber material in styrene and polymerizing the system. The blend contains not only rubber and polystyrene (PS), but also a graft polymer because of the attachment of short polystyrene side chains to the rubber molecules. The toughness of this material was markedly improved compared to that of the unmodified PS. A technology based on bulk polymerization [26] has been widely used; the concentrated emulsion polymerization method employed by us, however, allows one to obtain rubber toughened latexes.

Styrene-butadiene-styrene three-block copolymer (SBS) was used as a modifier to obtain toughened PS composites. Suitable amounts of SBS and azobisisobutyronitrile (AIBN) were introduced in styrene to generate a uniform styrene-rubber solution. An aqueous solution of sodium dodecyl sulphate (SDS) was placed in a flask. The styrene-rubber solution was introduced dropwise into the flask with vigorous stirring. The volume of the SDS aqueous solution was one-fourth of the styrene-rubber solution. The paste-like concentrated emulsion thus generated was transferred to a tube which, after being sealed with a rubber septum, was introduced into a water bath at 60 °C to perform the polymerization. In order to compare with the bulk polymerization, another styrene-rubber solution was transferred to another tube, introduced into an ultrasonic mixer, and kept at 60 °C for 6 h. After about 6 h the increased viscosity of the system made the ultrasonic mixing ineffective and the tube was transferred to a water bath at 60 °C. The material obtained by the concentrated emulsion method was transformed into a fine powder in a blender, washed in an extractor with methanol and dried in a vacuum oven. The product obtained by bulk polymerization was directly used in various testings.

The viscosities of the samples were measured with an Ubbelodhe viscometer at 25 °C, using tetrahydrofuran (THF) as solvent. They were used to calculate the intrinsic viscosities and the molecular weights (listed in Table 16) of the PS homopolymer present in the composite. High molecular weights of PS were achieved in both polymerizations, some higher than 10^6. This can be partly attributed to the presence of the rubber molecules in the monomer which, by increasing the viscosity of the system, decrease the biradical termination rate. This "gel effect" is additionally stimulated by the concentrated emulsion and, as a result, the molecular weights of the samples prepared via the concentrated emulsion pathway are higher than those prepared via bulk polymerization.

The tensile testing was conducted at room temperature, with an elongation speed of 50 mm/min. Figure 40 presents some of the stress–strain curves of the SBS modified PS composites. Each has a yield point typical of tough plastic materials. The yield strength decreases, however, as the rubber content increases. One may also notice that large deformations occur after stress softening, for small stress increases. The higher the rubber content, the larger the deformations. These stress–strain curves are very different from that of the unmodified PS, which is a very brittle material with no yield point and a very small deformation, below 4%.

Table 16. Intrinsic viscosities and molecular weights for SBS toughened composites

Sample[a]	Concentrated Emulsion			Bulk		
	Intrinsic Viscosity of Composite (ml g^{-1})	Intrinsic Viscosity of PS (ml g^{-1})	Mol. Weight of PS	Intrinsic Viscosity of Composite (ml g^{-1})	Intrinsic Viscosity of PS (ml g^{-1})	Mol. Weight of PS
SBS 0	–	132.0	423,000	–	99.6	287,000
SBS 10	250.1	267.6	1,121,000	199.9	212.5	816,000
SBS 15	214.7	235.0	937,000	160.2	173.1	615,000
SBS 20	196.4	220.4	858,000	125.4	135.7	436,000
SBS 25	178.3	204.1	771,000	101.3	108.1	321,000
SBS 30	137.3	155.9	532,000	93.0	98.7	283,000

[a] SBSx (with x = 0, 10, 15, 20, 25, 30) stands for wt. parts of SBS to 100 wt. parts of styrene

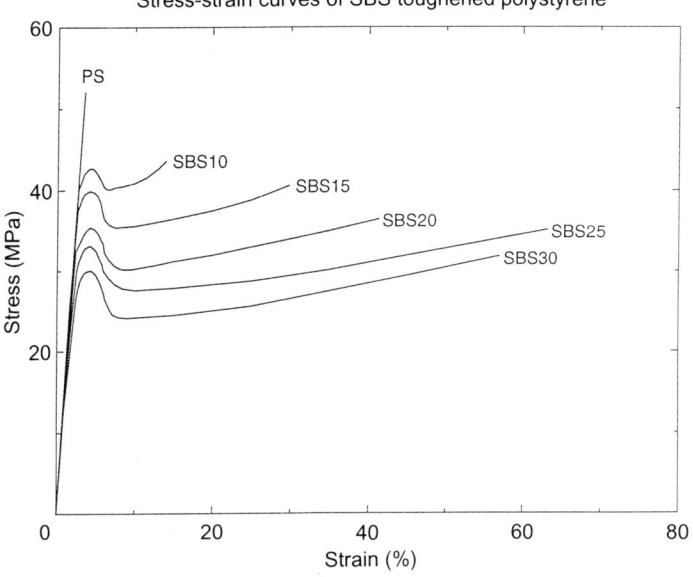

Fig. 40. Stress–strain curves of SBS toughened PS composites prepared via the concentration emulsion pathway. SBSx (with x = 10, 15, 20, 25 and 30) stands for wt. parts of SBS to 100 wt. parts of styrene

The tensile strength and elongation at the break point are listed in Table 17. The results for both types of polymerization are comparable and have similar trends: the higher the content of rubber, the higher the elongation and the lower the tensile strength. The toughness, calculated as the integral of the stress–strain curve, has similar trends: the greater the amount of rubber blended, the higher the toughness. This occurs because the flexible rubber segments impede the compact packing of the stiff PS segments; hence, the segments of PS can change

Table 17. Mechanical properties of SBS toughened composites

Sample[a]	Concentrated Emulsion			Bulk		
	Tensile Strength (MPa)	Elongation (%)	Break Energy (MJ m^{-3})	Tensile Strength (MPa)	Elongation (%)	Break Energy (MJ m^{-3})
SBS 10	43.6	14.0	5.2	56.5	8.7	4.4
SBS 15	40.6	29.8	10.9	51.1	26.0	12.6
SBS 20	36.4	41.3	15.0	47.5	34.0	16.2
SBS 25	35.1	63.2	26.6	41.0	61.8	27.9
SBS 30	31.8	56.9	23.5	36.5	64.9	28.4

[a] SBSx (with x = 10, 15, 20, 25, 30) stands for wt. parts of SBS to 100 wt. parts of styrene

their conformation (i.e., can have larger elongation) for less energy. The diluted density of segments, however, provides smaller inter-molecular interactions and hence a lower tensile strength. While the two polymerization methods provide comparable materials, the concentrated emulsion generates latexes that can be more easily processed in any desirable shape. Polystyrene toughened with rubber via the concentrated emulsion method can have an elongation which is 16 times larger than that of the ordinary polystyrene, the tensile strength being decreased by only 33%. When measured by the area under the stress–strain curves, the toughness of the toughened composites can be 15 times higher than that of ordinary polystyrene.

10.2 Preparation of MBSB Composites [28]

Extending the above methodology to a multi-component system, new composites, the MBSB composites, were prepared. They are based on the monomers methyl methacrylate (MMA), butyl methacrylate (BMA), styrene (ST) and the thermoplastic elastomer SBS. The copolymerization of ST, MMA and BMA generates a strong, tough and easily processable material, which is additionally toughened by the SBS molecules and by the grafting of the monomers onto the polybutadiene segments. The MBSB composites possess stress–strain curves similar to those of the other rubber toughened PS. Their mechanical properties are, however, much better than those of most styrene-based materials already produced. The latter have tensile strengths between 30–50 MPa and elongations at the break point of 40–20% [30], the higher tensile strength being associated with the smaller elongation. As shown in Table 18, the MBSB composites possess tensile strengths of 40–60 MPa and elongations greater than 50%. The effects of various weight ratios of different monomers are listed in Table 18. The larger the MMA/BMA ratio, the higher the tensile strength and the lower the elongation. When the ST/MMA ratio changes, the tensile strength remains

Table 18. The tensile properties of the MBSB composites

Wt. ratio[a] St/MMA/BMA	Concentrated Emulsion		Bulk Polymerization	
	Tensile Strength (MPa)	Elongation (%)	Tensile Strength (MPa)	Elongation (%)
Constant wt. fraction of Styrene = 0.45				
5/5/0	57.4	16.3	56.5	18.5
5/3/2	47.3	37.3	47.8	45.0
5/2.5/2.5	45.3	59.7	43.8	66.9
5/2/3	45.0	100.4	43.0	106.4
5/1/4	43.6	120.7	42.6	128.8
5/0/5	38.0	131.8	36.5	155.4
Constant wt. fraction of MMA = 0.18				
7/2/1	51.4	25.2	49.5	32.5
6/2/2	49.0	34.7	46.4	38.9
5/2/3	45.0	100.4	43.0	106.4
3/2/5	45.7	109.0	42.4	120.3
Constant wt. fraction of BMA = 0.27				
7/0/3	48.7	101.8	45.9	107.2
6/1/3	47.6	85.4	44.1	90.4
5/2/3	45.0	100.4	43.0	106.4
4/3/3	46.4	59.9	44.0	64.4

[a] The system also contains 20 g SBS per 100 ml monomer mixture

almost constant, even though the tensile strengths of MMA and styrene homopolymers are about 75 and 52 MPa, respectively. The elongation, however, decreases as the ST/BMA ratio increases. In general, the toughness of the composite decreases as the contents of MMA and styrene become larger. However, as the BMA content increases, the toughness passes through a maximum. Hence a proper amount of BMA can achieve optimum toughness.

In general, the tensile strength of the composites based on concentrated emulsions is higher than that based on bulk polymerization; the elongation at the break point and the toughness are, however, a little lower.

10.3 Poly(vinylidene chloride)/Poly(butyl methacrylate) Composites Prepared via the Concentrated Emulsion Pathway [29]

The copolymers of vinylidene chloride (VDC) and BMA have high flexibility. Their tensile strength and toughness can be further improved by including proper amounts of VDC and BMA homopolymers. The melt mixing method cannot be used because the melting point of PVDC is high (nearly 200 °C) [31]. The inclusion of the homopolymers via the concentrated emulsion method can, however, be easily achieved. Three preparation procedures were employed to obtain PVDC/PBMA composites. Because the boiling-point of VDC is 32 °C and the AIBN initiator requires 60 °C [32] to become sufficiently active, a redox system (ammonium persulfate, sodium metabisulfite and ferrous sulfate) [33] dissolved in the water phase was used to catalyze the polymerization through the oil-water interface.

Procedure 1 [29, 34] – first, separate concentrated emulsions of the two monomers, using an SDS aqueous solution containing the redox system as the continuous phase, were prepared. Each of the two concentrated emulsions was partially-polymerized at 45 °C until a 5% conversion was achieved. Subsequently the two concentrated emulsions were mixed. The mixture thus obtained was further heated at 45 °C to complete the polymerization. The obtained composites were washed with methanol and dried under vacuum at room temperature.

Procedure 2 [29] – PBMA was dissolved in VDC monomer to form a solution of 2 wt%. This solution replaced the VDC monomer of Procedure 1; the other steps have been as in Procedure 1.

Procedure 3 [29] – the preparation and partial-polymerization of a concentrated emulsion of VDC was the same as in Procedure 1. The BMA monomer containing the redox system was introduced dropwise at room temperature into the partially polymerized VDC emulsion. Because of the presence of water and surfactant in the emulsion, the system remained a concentrated emulsion after the addition of the BMA monomer. This concentrated emulsion was introduced into a water bath at 45 °C to complete the polymerization.

The samples prepared by the three procedures have been investigated by differential scanning calorimetry (DSC). Each sample was heated twice from -50 to 210 °C, with a heating rate of 10 °C min^{-1}; after the first heating, the sample was cooled quickly to a temperature below -50 °C and then heated for the second time. The DSC diagrams show that each composite has two glass transition temperatures, denoted as T_{g1} and T_{g2}, and a sharp melting point T_m. In contrast, the copolymers have only one T_g and a diffuse T_m. The T_{g1} s of the composites and copolymers are located in the narrow range of 20–36 °C, which may represent the glass transitions of both the BMA homopolymer and of the VDC/BMA copolymer type 1 (without long VDC sequences). The T_{g2} s are in the vicinity of 80 °C and are absent during the second heating. They are probably due to the crystallization of the long sequences of VDC units of the copolymer molecules (this VDC/BMA copolymer with long VDC sequences will be denoted as copolymer type 2). Since the long sequences crystallize, the motion of the BMA segments between two successive VDC crystals is rather restricted, and hence a T_{g2} appears at a higher temperature. Since the time between the first and second heating is too short for the VDC sequences to recrystallize, the chains of type 2 copolymer remain in an amorphous state during the second heating and the transition at the temperature T_{g2} is no longer present. The copolymer samples do not exhibit a T_{g2}.

The stress–strain curves, obtained with an elongation speed of 50 mm min^{-1}, are presented in Fig. 41, and the tensile properties in Figs. 42 and 43. The stress–strain curves show that the samples prepared by Procedures

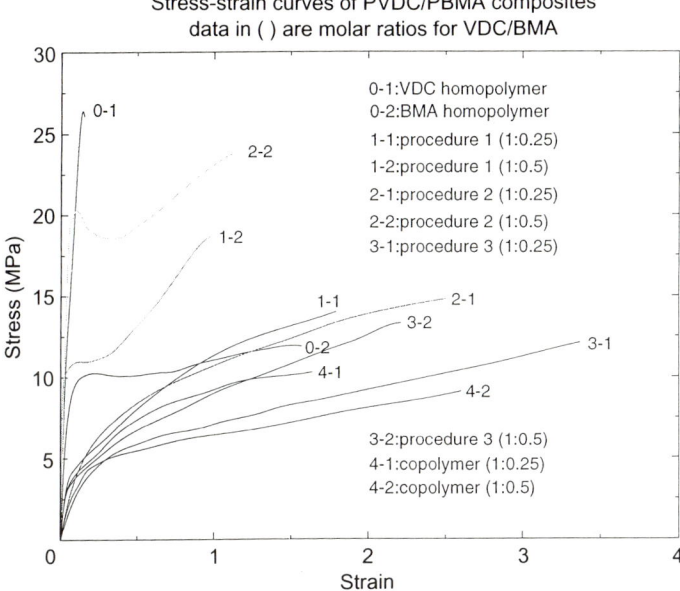

Fig. 41. Stress–strain curves of the PVDC/PBMA composites. The ratio in parenthesis represents the VDC/BMA molar ratio

0–1: VDC homopolymer   0–2: BMA homopolymer
1–1: procedure 1 (1:0.25)   1–2: procedure 1 (1:0.5)
2–1: procedure 2 (1:0.25)   2–2: procedure 2 (1:0.5)
3–1: procedure 3 (1:0.25)   3–2: procedure 3 (1:0.5)
4–1: copolymer (1:0.25)   4–2: copolymer (1:0.5)

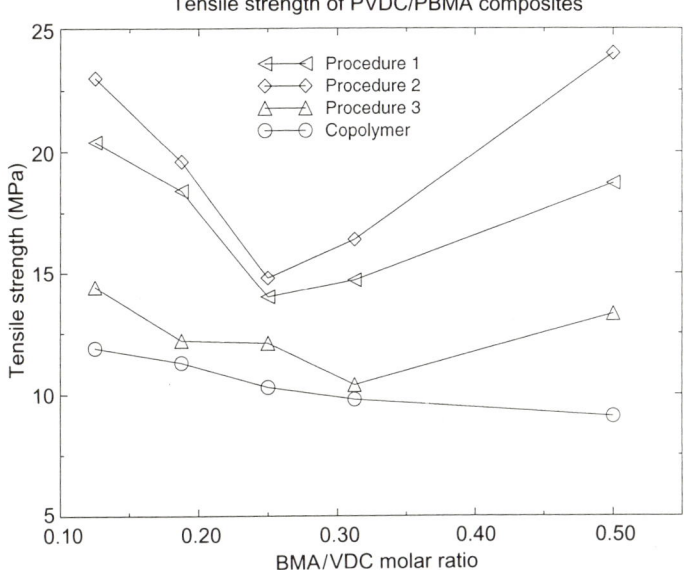

Fig. 42. Tensile strength of PVDC/PBMA composites

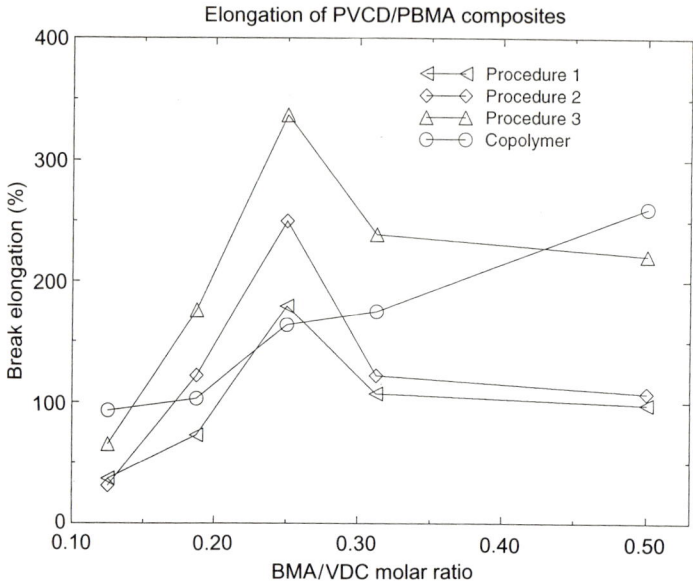

Fig. 43. Elongation of PVCD/PBMA composites

1 and 2 exhibit plastic behaviour, while those prepared by Procedure 3 and by copolymerization exhibit rubber-like behavior. For copolymers, the greater is the amount of BMA monomer, the higher the elongation (because the lateral butyl groups impede a compact arrangement of the polymer molecules) and the smaller the tensile strength (because the interactions are weakened). For composites, the tensile strength passes through a minimum and the elongation through a maximum as the molar ratio of VDC/BMA monomer increases. This can be explained as follows. The VDC homopolymer has a crystalline structure and hence a high tensile strength and a low elongation. At low BMA contents, copolymers form and the degree of crystallinity decreases. As a result, the tensile strength of the composite decreases and the elongation increases. At moderate contents of BMA, the VDC/BMA copolymer molecules acquire a large number of VDC segments, which are long enough to generate crystallized domains that are interlinked via BMA, or VDC/BMA type 1 segments. The formation of the interlinked domains is responsible for the increase of the tensile strength and the decrease of the elongation with increasing amount of BMA. Obviously, at large amounts of BMA (outside the range employed in the present experiments), the tensile strength should again decrease and the elongation increase with increasing amount of BMA. The concentrated emulsion pathway ensures not only suitable tensile strengths and elongations (Figs. 42 and 43), but also a suitable toughness (Table 19). The comparison with the copolymer (Table 19) indicates that the composites are much tougher.

Table 19. Work needed to break the samples (MJ m^{-3})

Mol Ratio (VDC/BMA)	Procedure 1	Procedure 2	Procedure 3	Copolymer
1:0.25	12.7	18.5	19.0	10.7
1:0.5	18.0	18.0	19.5	15.5

Comparing the results of Procedures 1 and 2 (Figs. 42 and 43) one can observe that the pre-dissolution of a small amount of PBMA into the VDC monomer has favorable effects for both the tensile strength and elongation.

The above results show that the methods based on concentrated emulsions can be employed to produce tough materials either by polymerization (copolymerization) and grafting, or by mixing partially formed latexes and polymerization. While the composites prepared via the concentrated emulsion polymerization exhibit comparable toughness and tensile properties to those prepared via the bulk polymerization, the concentrated emulsion methodology has the advantage of providing a fine powder, which can be easily employed in many applications. The bulk polymerization produces a bulk material which must be further pelletized with an extruder. For the method involving mixing of partially formed latexes, the concentrated emulsion has the advantages of higher yield and greater ease of handling than the conventional one.

11 Encapsulation of Solid Particles by the Concentrated Emulsion Polymerization Method [35]

Encapsulation is a process in which tiny particles or droplets are covered by a coating [36–43]. Its role is either to isolate the active ingredient or to control the rate by which it leaves the capsule. As examples for the first case, one can mention the isolation of vitamins from oxygen or of a reactive core from chemical attack, and for the second case, the control of the rate of release of drugs or pesticides. Numerous encapsulation techniques have been suggested. Most of the encapsulations of the active solid materials have been carried out for solids larger than 1 μm.

In this section the concentrated emulsion polymerization method is employed to encapsulate submicron inorganic powders. In a first step, a stable colloidal dispersion of the powder in an aqueous solution of a monomer containing an appropriate dispersant and a suitable initiator was prepared. This colloidal dispersion was subsequently employed as the dispersed phase of a concentrated emulsion whose continuous phase, decane, contained a surfactant.

The role of the surfactant is to stabilize the gel-like concentrated emulsion. Upon heating at 40 °C, polymerization took place and the solid particles were encapsulated in the polymer.

The experimental details are as follows. A small amount of decane containing Span 80 was placed in a flask equipped with a mechanical stirrer, an addition funnel, and a nitrogen inlet. Then a colloidal dispersion was prepared by dispersing the solid powder in an aqueous solution of acrylamide, N,N'-methylene bisacrylamide (crosslinking agent), and Triton X-45 (dispersant) under stirring. Subsequently, potassium persulfate (initiator) was added to the system. The preparation of the concentrated emulsion was carried out at room temperature by dropwise addition of the colloidal dispersion to the continuous phase within about 10 min under a nitrogen atmosphere. Polymerization was carried out in a water bath at 40 °C under a nitrogen stream for 6 h.

Scanning electron micrographs of alumina powders dispersed in water in the absence and presence of dispersant show that the agglomeration of the solid particles decreases tremendously in the presence of the dispersant. The agglomerates are larger than 10 μm in the absence of dispersant and smaller by more than one order of magnitude in its presence.

Figure 44 presents a scanning electron micrograph of a capsule in which alumina particles are encapsulated in crosslinked polyacrylamide. Table 20 lists under PLA1 the amounts of the components involved in the preparation of these capsules. The capsules have a polyhedral shape and their sizes are larger (around 5 μm) and more uniform than the polymer latexes free of solid particles. Some of the cells of the gel coalesce during polymerization, forming bulk phases. As a result, some unencapsulated solid particles were also observed.

Very fine solid particles, namely fumed silica, were also encapsulated via the concentrated emulsion polymerization method. The amounts of the components involved are listed under PLS1 in Table 21. The PLS1 capsules range in size from 1.0 to 1.5 μm.

12 Concentrated Emulsion Polymerization Pathway to Hydrophobic and Hydrophilic Microsponge Molecular Reservoirs [44]

Porous polymer particles are used in chromatography, ion exchange, as reactive polymer matrixes, and more recently in controlled release of drugs [45]. The synthesis of these particles is, in general, based on the suspension polymerization of a dispersed phase that consists of a monomer, a cross-linking agent, an initiator, and a suitable inert solvent that functions as porogen [46]. The size of the particles depends on the suspending medium as well as on the nature and amount of dispersant it contains. The porosity of the particles is controlled by the volume fraction of the porogen and by the concentration of the cross-linking

Concentrated Emulsion Polymerization 51

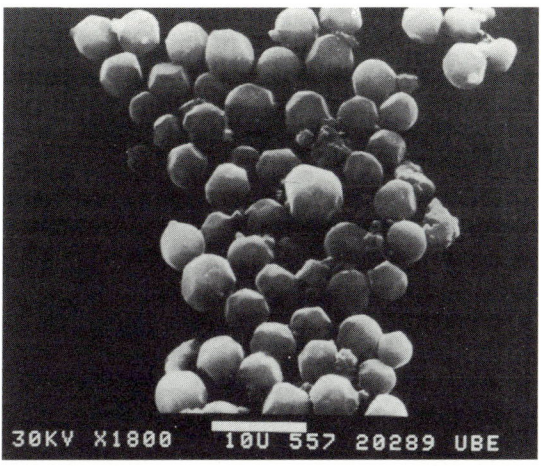

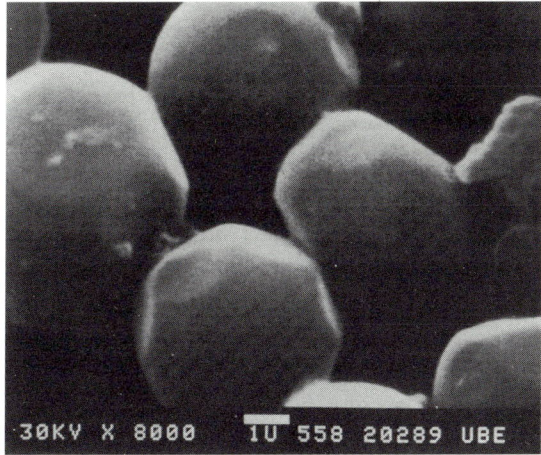

Fig. 44. Scanning electron micrographs at two magnifications of capsules of alumina particles encapsulated in crosslinked polyacrylamide with the composition PLA1 (Table 20)

Table 20. Representative compositions in the preparation of alumina capsules

	PLA 1
Dispersed phase	
Acrylamide (g)	5
N,N'-methylene bisacrylamide (g)	0.5
γ-alumina (g)	1.5
Dispersant (Triton X-45) (g)	2
Initiator (sodium persulfate) (g g^{-1} acrylamide)	0.01
Water (g)	20
Continuous phase	
Decane (ml)	3
Surfactant (sorbitan monooleate) (ml)	1.5

Table 21. Representative compositions in the preparation of silica capsules

	PLS 1
Dispersed phase	
Acrylamide (g)	5
N,N'-methylene bisacrylamide (g)	0.5
Fumed silica (g)	0.75
Dispersant (Triton X-45) (g)	2
Initiator (sodium persulphate) (g g^{-1} acrylamide)	0.01
Water (g)	20
Continuous phase	
Decane (ml)	3
Surfactant (sorbitan monooleate) (ml)	1.5

agent [47]. In the present section, a novel method for the preparation of porous polymer particles which is based on the polymerization of concentrated emulsions is presented. In contrast to suspension polymerization, the concentrated emulsion method employs a small volume fraction of continuous phase and this allows easier surface modifications [48].

Four kinds of polymeric beads have been prepared [44a]: (1) porous highly cross-linked hydrophobic polystyrene particles; (2) lightly cross-linked hydrophilic polyacrylamide particles; (3) porous nutshells of highly cross-linked polystyrene encapsulating sparse poly (VBC) matrixes; (4) hydrophilic particles with lightly cross-linked polyacrylamide nutshells encapsulating loosely entangled poly (EO) molecules. These particles can be used as reservoirs for chemically active species and medicines. We employed such particles for trapping carbonyl complexes or for immobilizing the enzyme lipase.

Cross-linked polystyrene porous particles (with 21 mol% DVB) have been prepared by the concentrated emulsion polymerization method, using either toluene or decane as the porogen and an aqueous solution of SDS as the continuous phase. Since toluene is a good solvent for polystyrene while decane is a "nonsolvent", the morphologies obtained in the two cases were different. The particles based on toluene (with a volume fraction of dispersed phase of 78%) have very small pores which could not be detected in the SEM pictures. The pore size distribution, which has sizes between 20 and 50 Å and was determined with an adsorption analyzer, almost coincides with that in a previous study [49] in which porous polystyrene beads have been prepared by suspension polymerization. In contrast, the porous particles based on decane have pore sizes as large as 0.1–0.3 μm, which could be detected in the SEM pictures [44a], and also larger surface areas (47 m^2 g^{-1}) than those based on toluene (25 m^2 g^{-1}). The main difference between the concentrated emulsion polymerization and the suspension polymerization consists of the much smaller volume fraction of continuous phase used in the former procedure. The gel-like emulsion that constitutes the precursor in the former case contains polyhedral cells separated by thin films of continuous phase. The polymerization of the cells does not appreciably change their size, which is in the range 0.1–10 µm. In suspension polymerization the sizes are larger, in the range 3–300 µm [45]. The fact that the volume fraction of the continuous phase is very small in the concentrated emulsion method is particularly advantageous in the preparation of porous nutshell particles presented below.

A two-step concentrated emulsion polymerization procedure was employed to prepare porous nutshell particles. In the first step, a gel-like concentrated emulsion of a mixture of solvent/non-solvent for the polymer (as porogen), containing small amounts of VBC and DVB and the initiator AIBN, was dispersed in water and polymerized at 40 °C for 16 h. In the second step, a mixture of styrene and DVB was introduced into the polymerized gel-like emulsion, and the resulting system was polymerized for three days. During the first polymerization, the cross-linked poly(VBC) matrix is swollen by the solvent (toluene), and a nucleus is generated surrounded by a liquid layer containing the nonsolvent (decane) as the main ingredient. The mixture of styrene-DVB, introduced in the second step, mixes with decane but cannot penetrate into the already polymerized nucleus. Its polymerization generates a porous nutshell of crosslinked polystyrene surrounding a "void" filled with a sparse poly(VBC) matrix (this mechanism is depicted in Fig. 45). Since the volume fraction of the continuous phase is small, it is easier for the mixture of styrene-DVB to surround the prepolymerized particles than when the volume fraction is large.

SEM pictures [44a] show that the nutshell contains marble-like pieces, with submicron size pores among them, which are a result of the non-solvent effect of decane on polystyrene.

A suspension of nutshell particles in DMF was quaternized with tributylphosphine. This converted the encapsulated poly(VBC) into poly[(vinylbenzyl)-tributylphosphonium chloride]. The EDS experiments indicated a low surface concentration of phosphorus in the quaternized samples. Since the elemental analysis indicated a high content of phosphorus in all the samples, one can infer that the quaternary onium cations are hidden inside the polystyrene "cages".

Molecular reservoirs consisting of a porous polydivinylbenzene (PDVB) nutshell surrounding polytriethylvinylbenzyl ammonium chloride (PEVAC) chains were employed to immobilize the homogeneous catalysts $[Co(CO)_4]^-$ or $[Rh(CO)_2I_2]^-$ [44b]. The molecular reservoirs were prepared by the successive polymerization of vinylbenzyl chloride (VBC) and divinylbenzene (DVB) in a concentrated emulsion, followed by the quaternization of the encapsulated VBC chains. The immobilization was achieved via the formation of ion-pairs between the ammonium cations and the metal carbonyl anions. Two kinds of polymer-bound catalysts dispersed in water were used to catalyze the synthesis of two derivatives of styrene oxide in the presence of CO. It was found that when PDVB(PEVAC)–$[Co(CO)_4]^-$ was used, 2,5-dihydro-2-oxo-3-hydroxy-4-phenylfuran was the main product, as in the homogeneous case. When, however, PDVB(PEVAC)–$[Rh(CO)_2I_2]^-$ was employed, 2-phenyl-2-(α-styryoxyl)-ethanol was obtained as the main product; in the homogeneous case the

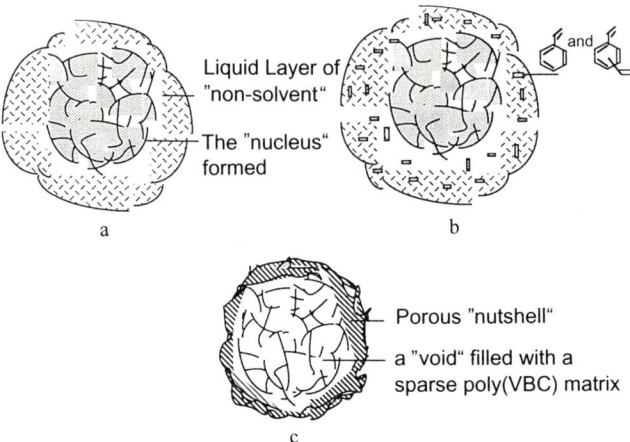

Fig. 45A–C. Schematic illustration of the process for generating porous nutshell particles: **A** cell containing a nucleus; **B** polymerization of styrene/DVB in the non-solvent layer; **C** formation of the porous nutshell particle

selectivity was much lower. Both polymer-bound catalysts were used in five successive cycles at 200 psi CO and 95–98 °C, and the conversion of styrene oxide was almost quantitative. The cocatalyst methyl iodide was introduced before the first cycle only. It was found that 72% of the initial Co and 94% of the initial Rh were still present in the molecular reservoirs after the five cycles. In addition to immobilizing the homogeneous catalyst, the molecular reservoir PDVB(PEVAC) also possesses some adsorption capability for the organic reactant molecules because of its hydrophobic porous PDVB shell. This hydrophobic adsorption capability allows the styrene oxide molecules to approach the encapsulated catalytically active sites.

Porous hydrophilic particles have previously been prepared [53] via suspension polymerization. A concentrated emulsion of water (containing a hydrophilic monomer (acrylamide or sodium acrylate) and a cross-linking agent) in oil containing Span-80 as emulsifier was employed by us to prepare hydrophilic particles.

SEM pictures [44a] have shown that the hydrophilic porous spherical particles of polyacrylamide are of micron size and have a smooth surface. The hydrophilic particles can be used as biocompatible substrates for the solid-phase synthesis of peptides [54] as well as for immobilization of enzymes. We determined, for instance, that these hydrophilic particles can adsorb 68 mg g^{-1} of lipase from an aqueous solution containing 0.5 mg ml^{-1} of lipase.

If first a concentrated emulsion of w/o in which the dispersed medium is an aqueous solution of poly(ethylene oxide) is prepared, followed by the addition of acrylamide and N,N'-methylenebisacrylamide and polymerization, a polyacrylamide shell that encapsulates the poly(ethylene oxide) can be generated [44]. This reservoir could be used in the controlled release of drugs. Indeed, if a biodegradable polymer (which is often water soluble) and drug molecules are encapsulated into the polyacrylamide shell particles, the releasing process can be controlled via the degrading action of a specific enzyme present in the body.

13 Other Applications of Concentrated Emulsions

In summary, concentrated emulsions have been used to prepare latexes or composites, to encapsulate solid particles, for polymer blending and to generate "molecular" reservoirs. The mechanical properties could be controlled by combining suitable monomers in the latexes. The concentrated emulsion pathway was also employed in our laboratory in other directions than those presented in this review, and we would like to mention them in this final section without details, but with suitable references.

13.1 Selective Composite Membranes

Selective composite membranes [55–61] have been prepared by choosing as the dispersed phase of the concentrated emulsion a system that contains a monomer (or monomers) leading to a polymer (polymers) which swells in those components of the mixture for which the membrane should be selective. The continuous phase, on the other hand, was selected to yield a polymer that did not swell in any of the components of the mixture, its role being to maintain the integrity of the membrane. As examples, composite membranes containing polystyrene as the dispersed phase and polyacrylamide as the continuous phase have been used to separate toluene from cyclohexane, and those containing acrylamide as the dispersed phase and crosslinked polystyrene as the continuous phase to separate water from ethanol. Simple considerations based on chemical intuition could be used to design the membranes. The mechanical properties of the membranes could be tremendously improved by including in one of the phases an additional suitable monomer.

13.2 Conductive Composite Polymers

Concentrated emulsions have been used to prepare conductive composite polymers, either by polymerizing pyrrole or aniline within the pores of a porous material [62–64] or by preparing conductive lattices [65–67].

13.3 Polymer Supported Catalytic Groups

Polymer supported quaternary onium salts [68] and polymer supported catalysts containing phosphorus palladium complexes and quaternary onium groups [69] have been prepared via the concentrated emulsion method. The first catalysts have been employed as phase transfer catalysts in the alkylation of isopropylidene malonate [68], and the second catalysts have been employed for the vinylation reaction [69]. In addition, crosslinked polystyrene latexes with surface bound quaternary ammonium groups have been prepared by a two-step method [70]. In the first step, the functional monomer vinylbenzene chloride (VBC) was introduced into a partially polymerized concentrated emulsion of styrene and divinylbenzene (ST/DVB) in water and complete polymerization was carried out. In the second step, the poly(ST/DVB) particles with surface bound benzylchloride groups were quaternized with various amounts of hydrophilic quaternary onium groups. The dual-surface characteristics due to the hydrophobic polystyrene and to the hydrophilic bound onium cations could be varied by changing the duration of partial polymerization of ST/DVB and the amount of VBC. The phase transfer catalysts thus prepared were used in the alkylation reaction of isopropylidene malonate. The conversion of isopropylidene malonate increased with increasing partial polymerization time, while the ratio of the mono-to dialkylation products increased with decreasing partial polymerization time of the concentrated emulsion. This can be explained as follows. When the partial polymerization time is short, part of the VBC copolymerizes with styrene and another part homopolymerizes and is grafted to the surface of the particles. When the partial polymerization time is long, most of the VBC homopolymerizes and the chains are grafted to the surface of the particles. The ammonium groups to which the benzyl chloride units are converted via quarternization are more accessible, for steric reasons, to the isopropylidene malonate and the monoalkylated product when they are part of the grafted chains rather than part of the copolymer. The hydrophilic recognition by the grafted chains of the isopropylidene malonate and of the monoalkylated product is hence responsible for the changes in activity and selectivity.

13.4 Polymer Substrates for the Immobilization of Enzymes and Cells

Concentrated emulsions were also used to prepare substrates for the immobilization of enzymes and cells. Lipase was immobilized in hydrophobic porous polymers and the system employed in the hydrolysis of triacylglycerides [71]. Cells of *Phanerochaete Chrysosporium* were immobilized on porous poly(styrene-divinylbenzene) carrier and used for the degrading of 2-chlorophenol [72]. The substrates were prepared as in Ref. 73.

14 References

1. Lissant KJ (1966) J Colloid Interface Sci 22: 492
2. Lissant KJ, Mayhan KG (1972) J Colloid Interface Sci 42: 201
3. Lissant KJ, Peace BW, Wu SH, Mayhan KG (1974) J Colloid Interface Sci 47: 416
4. Princen HM, Aronson MD, Moser JC (1980) J Colloid Interface Sci 75: 246

5. Princen HM (1983) J Colloid Interface Sci 91: 160; (1985) J Colloid Interface Sci 105: 150
6. Princen HM, Kiss AD (1986) J Colloid Interface Sci 112: 427
7. Kraynick AM, Hansen MG (1986) J Rheology 30: 409
8a. Ruckenstein E, Ebert G, Platz G (1989) J Colloid Interface Sci 133: 432
8b. Bhakta A, Ruckenstein E (1995) Langmuir 11: 1486
8c. Bhakta A, Ruckenstein E (1995) Langmuir 11: 4642
9. Ruckenstein E, Park JS (1992) Polymer 33: 405
10. Ruckenstein E, Kim KJ (1988) J Applied Polym Sci 36: 907
11. Kim KJ, Ruckenstein E (1988) Makromol Chem Rapid Commun 9: 285
12. Ruckenstein E, Park JS (1988) J Polym Sci (C) Polym Lett 26: 529
13. Becher P (1965) Emulsion, Theory and Practice. Reinhold, New York
14. Smith AL (1976) Theory and Practice of Emulsion Technology, Academic Press, New York Ch 19
15. Chen HH, Ruckenstein E (1990) J Colloid Interface Sci 138: 473
16. Chen HH, Ruckenstein E (1991) J Colloid Interface Sci 145: 260
17. Rosen MJ (1989) Surfactants and Interfacial Phenomena, 2nd Edn Wiley, New York
18. Schick MJ (1967) Nonionic Surfactants, Marcel Dekker, New York Vol 1
19. Sun F, Ruckenstein E (1993) J Applied Polymer Sci 48: 1773
20. Dimonie MV, Boghina CM, Marinescu MM, Cincu CI, Oprescu CG (1982) Eur Polymer J 18: 639
21. Kurata M, Tsunashima Y, Iwama M, Kamada K (1975) in Polymer Handbook (Brandup J, Immergut EH Eds) 2nd edition Wiley-Interscience, New York p IV: 9
22. Ruckenstein E, Kim JK (1989) J Polym Sci Part A Polym Chem 27: 4375
23. Ruckenstein E, Sun F (1992) J Appl Polym Sci 46: 1271
24. Ruckenstein E, Park JS (1989) Chem Mat 1: 343
25. Sperling LH (1991) Polym Mater Sci Eng 65: 80
26. Manson JA, Sperling LH (1976) Polymer Blends and Composites, Plenum Press, New York
27. Ruckenstein E, Li H (1994) J Apply Polym Sci 52: 1949
28. Ruckenstein E, Li H (1994) J Apply Polym Sci 54: 561
29. Ruckenstein E, Li H (1994) Polymer 35: 4343
30. Brydson JA (1979) Plastic Materials, 3rd Ed newnes-Butterworths, London
31. Zutty NL, Whitworth SJ (1964) J Polymer Sci B 2: 709
32. Brandrup J, Immergut EH (1989) Ed Polymer Handbook, 3rd Ed John Wiley & Sons, New York
33. Lee KC, El-Aasser MS, Vanderhoff JW (1991) J Appl Polym Sci 42: 3133
34. Ruckenstein E, Park JS (1990) Polymer 31: 2397
35. Park JS, Ruckenstein E (1990) Polymer 31: 175
36. Nixon JR (1976) Microencapsulation, Marcel Dekker Inc, New York
37. Das KG (1983) Controlled Release Technology, Wiley, Interscience, New York
38. Bakan JA (1980) Microencapsulation Using Coacervation/phase Separation Techniques in Controlled Release Technologies: Method, Theory, and Applications (Ed Kydoneius AF), CRC Press, Florida
39. Koestler RC (1980) Microencapsulation by Interfacial Polymerization Techniques – Agricultural Applications in Controlled Release Technologies: Methods, Theory, and Applications (Ed Kydoneius AF), CRC Press, Florida
40. Brynko C (1961) US Patent 2969330
41. Goldenhersh KK, Huang W, Manson NS, Sparks RE (1976) Kidney Int 10: 251
42. Hasegawa M, Arai K, Saito S (1987) J Polym Sci Polym Chem Edn 25: 3117
43. Hasegawa M, Arai K, Saito S (1987) J Appl Polym Sci 33: 411
44a. Ruckenstein E, Hong L (1992) Chem Mat 4: 1032
44b. Hong L, Ruckenstein E (1995) J of Molecular Catalysis A: Chemical 101: 115
45. Eury R, Patel R, Longe K, Cheng T, Nacht S (1992) Chemtech Jan 42
46. Sherrington DC (1982) In Macromolecular Syntheses; Pearce EM, Ed Wiley, New York Vol 8 p 30
47. Howard GJ, Midgley CA (1981) J Appl Polym Sci 26: 3845
48. Ruckenstein E, Hong L (1992) Chem Mater 4: 122
49. Guyot A, Bartholin M (1982) Prog Polym Sci 8: 277
50. Alper H, Aroumainian H, Petrignani JF, Manul J (1985) J Chem Soc Chem Commun 340
51. Edgell WF, Lyford IV J (1970) J Chem Phys 52: 4329
52 (a) Alper H (1991) Aldrichimica 24: 3 (b) Brunet J (1990) J Chem Rev 1041
53. Mauz O, Sauber K, Noetzel S (1985) US Patent 4,542,069

54. Bayer E (1991) E Angew Chem Int Ed Engl 30: 113
55. Ruckenstein E (1989) Colloid Polymer Sci 267: 792
56. Park JS, Ruckenstein E (1989) J Appl Polym Sci 38: 453
57. Ruckenstein E, Park JS (1990) J Appl Polym Sci 40: 213
58. Ruckenstein E, Chen HH (1991) J Appl Polym Sci 42: 2429
59. Ruckenstein E, Sun F (1993) J Membrane Sci 81: 191
60. Sun F, Ruckenstein E (1993) J Membrane Sci 85: 59
61. Sun F, Ruckenstein E (1994) J Membrane Sci 90: 275
62. Ruckenstein E, Park JS (1991) J Applied Polym Sci 42: 925
63. Ruckenstein E, Park JS (1991) Synth Met 44: 293
64. Ruckenstein E, Chen JH (1991) J Appl Polym Sci 43: 1209
65. Park JS, Ruckenstein E (1992) J Electronic Mat 21: 205
66. Ruckenstein E, Yang S (1993) Synth Met 53: 283
67. Yang S, Ruckenstein E (1993) Synth Met 60: 249
68. Hong L, Ruckenstein E (1992) Polymer 33: 1968
69. Hong L, Ruckenstein E (1991–92) Reactive Polymers 16: 181
70. Ruckenstein E, Hong L (1992) J Catalysis 136: 378
71. Ruckenstein E, Wang X (1993) Biotechnol Bioeng 42: 821
72. Ruckenstein E, Wang X (1994) Biotechnol Bioeng 44: 79
73. Barby D, Haq Z (1982) Eur Pat 0,060,138

Editor: J.L. Koenig
Received: Nov. 1995

N-Benzyl and *N*-Alkoxy Pyridinium Salts as Thermal and Photochemical Initiators for Cationic Polymerization

Yusuf Yagci[1] and Takeshi Endo[2]
[1] Istanbul Technical University, Department of Chemistry, Maslak, Istanbul 80626, Turkey
[2] Tokyo Institute of Technology, Research Laboratory of Resources Utilization, 4259 Nagatsuda-Cho, Midori-Ku, Yokohama, 226, Japan

Cationic polymerizations induced by thermally and photochemically latent *N*-benzyl and *N*-alkoxy pyridinium salts, respectively, are reviewed. *N*-Benzyl pyridinium salts with a wide range of substituents of phenyl, benzylic carbon and pyridine moiety act as thermally latent catalysts to initiate the cationic polymerization of various monomers. Their initiation activities were evaluated with the emphasis on the structure-activity relationship. The mechanisms of photoinitiation by direct and indirect sensitization of *N*-alkoxy pyridinium salts are presented. The indirect action can be based on electron transfer reactions between pyridinium salt and (a) photochemically generated free radicals, (b) photoexcited sensitizer, and (c) electron rich compounds in the photoexcited charge transfer complexes. *N*-Alkoxy pyridinium salts also participate in ascorbate assisted redox reactions to generate reactive species capable of initiating cationic polymerization. The application of pyridinium salts to the synthesis of block copolymers of monomers polymerizable with different mechanisms are described.

1	Introduction	61
2	*N*-Alkyl Pyridinium Salts as Thermally Latent Initiators for Cationic Polymerization	62
2.1	Synthesis	62
2.2	Polymerization	62
2.2.1	Effect of Substituents on Benzene Ring and Benzylic Carbon	64
2.2.2	Effect of Pyridine Moiety	65
2.3	Grafting onto Polystyrene Possessing *N*-Benzyl Pyridinium Salt Side Groups	68
3	*N*-Alkoxy Pyridinium Salts as Photoinitiators for Cationic Polymerization	68
3.1	Synthesis	69
3.2	Initiation of Cationic Polymerization	69
3.2.1	Direct Initiation	69
3.2.2	Indirect Initiation	71

- 3.3 Free Radical Polymerization by using *N*-Alkoxy Pyridinium Salts 80
- 3.4 Synthesis of Block Copolymers by Using *N*-Alkoxy Pyridinium Salts 80

4 Conclusions 84

5 References. 84

1 Introduction

Cationic polymerizaton of vinyl and cyclic monomers can be initiated by various initiators such as protonic and Lewis acids, carbocations, and trialkyloxonium salts [1]. General reactions for the initiation of the polymerization may be represented as shown below (Scheme 1).

In these processes, on mixing the monomer and initiator an exothermic polymerization ensues and low-temperature conditions are often required. Moreover, homogeneous mixtures of initiators and monomers cannot be achieved before polymerization commences. Therefore, initiators which induce polymerization by external stimulation such as photoirradiation and heating are extremely important in the control of the initiation step of cationic polymerization [2]. In this connection it should be pointed out that the polymers obtained by using externally stimulated initiators usually show broad molecular weight distribution. The initiating species are continuously generated by thermolysis or photolysis until all of the initiator is consumed.

During the past two decades, considerable attention has focused on several externally stimulated initiators, for example onium salts that undergo photolysis and thermolysis to initiate cationic polymerization. The major classes of externally stimulated initiators which can succesfully be applied to initiate cationic polymerization are collected in Table 1.

In this article we shall limit our report to *N*-benzyl- and *N*-alkoxy-pyridinium salts which have been of interest to both authors in recent years.

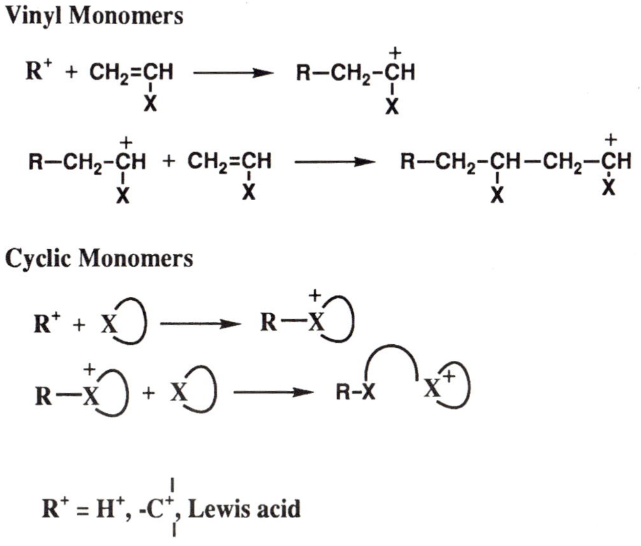

Vinyl Monomers

$$R^+ + CH_2=CH(X) \longrightarrow R-CH_2-\overset{+}{C}H(X)$$

$$R-CH_2-\overset{+}{C}H(X) + CH_2=CH(X) \longrightarrow R-CH_2-CH(X)-CH_2-\overset{+}{C}H(X)$$

Cyclic Monomers

$$R^+ + X\!\!\bigcirc \longrightarrow R-\overset{+}{X}\!\!\bigcirc$$

$$R-\overset{+}{X}\!\!\bigcirc + X\!\!\bigcirc \longrightarrow R-X\frown X^+\!\!\bigcirc$$

$$R^+ = H^+, -\overset{|}{\underset{|}{C}}{}^+, \text{Lewis acid}$$

Scheme 1

Table 1. The externally stimulated initiators for cationic polymerization

Initiator	The mode of stimulation	References
Diaryliodonium salts[a]	Photo	[3, 4]
Triarylsulphonium salts	Photo	[5, 6]
Benzylsulphonium salts	Photo, Thermal	[7–17]
Allylsulphonium salts	Photo, Thermal	[18]
Aryldiazonium salts	Photo, Thermal	[19, 20]
Phosphonium salts	Photo, Thermal	[21–23]
Benzylammonium salts	Thermal	[24–27]
Benzyl pyridinium salts	Thermal	[28–30]
Alkoxy pyridinium salts[a]	Photo	[31]
Iron arene complexes	Photo	[32, 33]
Sulphonyloxy ketones	Photo	[34, 35]
Silyl benzyl ethers	Photo	[35, 36]

[a] These salts may also be activated thermally by indirect methods [37–38]

2 N-Benzyl Pyridinium Salts as Thermally Latent Initiators for Cationic Polymerization

2.1 Synthesis

N-Alkyl pyridinium salts can readily be prepared by the reaction of benzyl chlorides with corresponding pyridines [28]. This procedures gives pyridinium salts possessing nucleophilic chloride anions. The synthesis of thermally latent initiator is completed by exchanging this anion for a less reactive nucleophilic one, e.g. SbF_6^-. Various salts of the following structures were synthesized by the proper choice of both benzyl chloride and pyridine and the spectral properties of the salts obtained are collected in Table 2 (Scheme 2). α-Methyl benzyl bromide was used in the case of **2e**, to avoid the Menscutting reaction which proceeded in parallel with the corresponding bromide, probably due to the steric hindrance of the methyl group [29]. The solvent was also changed to acetonitrile in order to prevent a back reaction caused by deposition of the product as precipitate.

2.2 Polymerization

N-Benzyl pyridinum salts were found to initiate thermally cationic polymerization of cyclic and vinyl monomers. The representative monomers with the following structures were successfully polymerized:

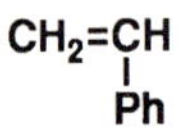

Styrene (St)	Glycidyl phenyl ether (GPE)	1,Phenyl-4-ethyl-2,6,7-trioxabicyclo [2.2.2]octane (BOEE)
$CH_2=CH$–Ph	$PhOCH_2CH$—CH_2 (epoxide)	Ph–C(–O–)$_3$–C–Et

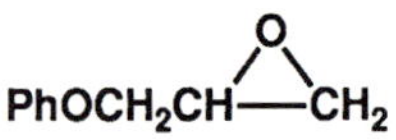

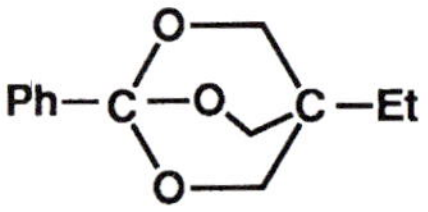

Table 2. H^1-NMR and IR spectral characteristics and melting points of various N-benzyl pyridinium salts

Pyridinium salt	^{1}H-NMR σ(ppm)	IR (cm^{-1})	mp (°C)	Ref.
2a	9.0–8.7(d), 8.7–8.45(m), 8.3–7.8(br), 7.5(s), 5.75(s)	1630, 1485, 1455, 760, 745, 700, 660	149.5–150.5	[24]
2b	8.98(d), 8.07(d), 7.47(s), 5.9(s), 2.72(s)	1645, 779, 736, 701, 653	149.5–150.5	[26]
2c	9.2–7.9(m), 7.47(s), 6.04(s), 2.97(s)	1634, 793, 744, 703, 653	149.5–150.0	[26]
2d	9.1–8.83(d), 8.55–8.15(br), 7.5(s)	1640, 1455, 760, 720, 700, 660	156–157	[24, 29]
2e	9.06–8.78(d), 8.43–8.1(br), 7.45(s), 6.33–5.87(q), 2.23–1.93(d)	1639, 1454, 767, 705, 659	108.5–109.5	[29]
2f	9.0–8.75(d), 8.45–8.15(br), 7.63–6.85(q), 5.81(s), 5.75(s)	1643, 1610, 1589, 1259, 1238, 1033, 771, 714, 655	157.5–158.5	[29]
2g	8.93(d), 8.3(d), 7.45(s), 5.75(s)	1640, 1600, 1460, 835, 660	147.0–147.5	[30]
2h			154–155	[30]
2i			120–124	[30]
2j			153	[30]
2k	9.5–8.4(br), 7.77–6.87(q), 6.2(s), 3.83(s)	1613, 1257, 1181, 784, 756, 711, 659	118–120	[20]
2l	10.5–8.1(m), 7.85–6.65(q) 6.37(s), 3.95(s), 3.74(s)	1744, 1639, 796, 744, 660	97–98.5	[26]

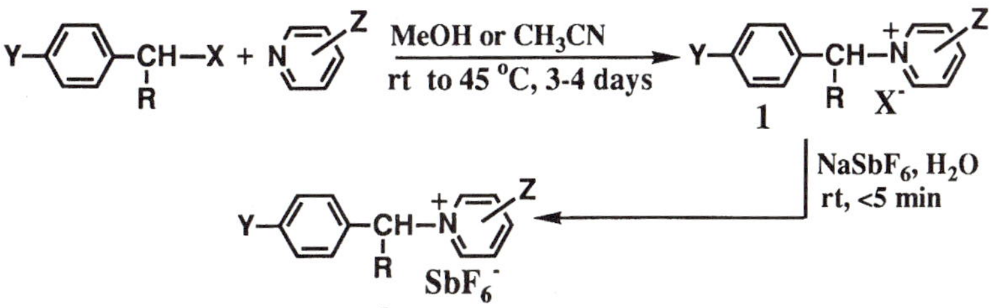

X:Cl

2a; Y:H, R:H, Z:H   2e; Y:H, R:CH$_3$, Z:p-CN   2i; Y:p-tC$_4$H$_5$, R:H, Z:p-CN

2b; Y:H, R:H, Z:p-CH$_3$ 2f; Y:p-OCH$_3$, R:H, Z:p-CN   2j; Y:H, R:H, Z:o-CN

2c; Y:H, R:H, Z:o-CH$_3$ 2g; Y:p-Cl, R:H, Z:p-CN   2k; Y:p-OCH$_3$, R:H, Z:o-CN

2d; Y:H, R:H, Z:p-CN   2h; Y:p-CH$_3$, R:H, Z:p-CN   2l; Y:p-OCH$_3$, R:H, Z:m-COOCH$_3$

Scheme 2

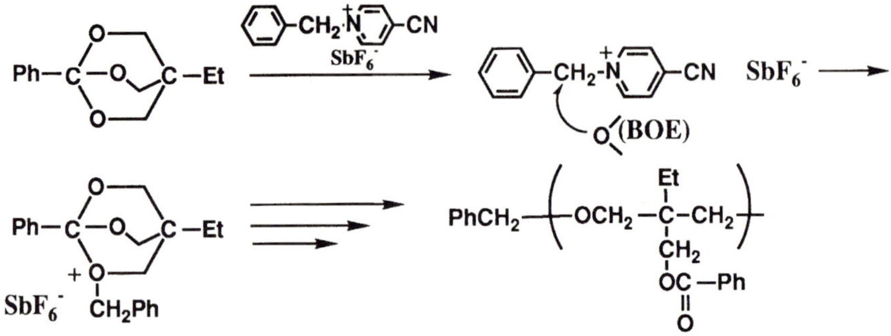

[Scheme 3]

Scheme 3

Benzyl pyridinum salts exhibit much better initiation activity than the aliphatic ammonium salts. In earlier investigations, on the basis of elemental analysis of the polymer obtained, the mechanism below was proposed for the initiation by pyridinium salts is shown in Scheme 3 [24].

2.2.1 Effect of Substituents on Benzene Ring and Benzylic Carbon

Although the p-CN substituted benzyl pyridinium salts **2d** is the most reactive initiator among the various quaternary ammonium salts, it requires rather high temperatures for a succesful polymerization. The initiation activity of these salts can be increased by introducing electron releasing groups onto the benzene ring and benzylic carbon [29]. Indeed, the α-methylbenzyl **2e** and p-methoxybenzyl **2f** derivatives exhibited much higher activity than the parent p-cyanopyridinium salt **2d** and thus polymerization may be conducted at much lower temperatures. The activity of these initiators was evaluated in the polymerization of GPE in which initiators are soluble at temperatures above 60 °C and polymerizations can be performed in homogeneous systems (Scheme 4).

In general, polymerizations with these salts are much more efficient than those with the unsubstituted salt. Considering that the structure of the propagating species and the rate of the polymerization (k_p) are expected to be the same in all three cases, the enhanced activity may be attributed to the stabilization of the benzyl cation by the substituents. Similar effects were observed with the benzylic sulphonium salts. However, more detailed studies [30] of polymerization and hydrolytic properties of various p-substituted

[Scheme 4]

Scheme 4

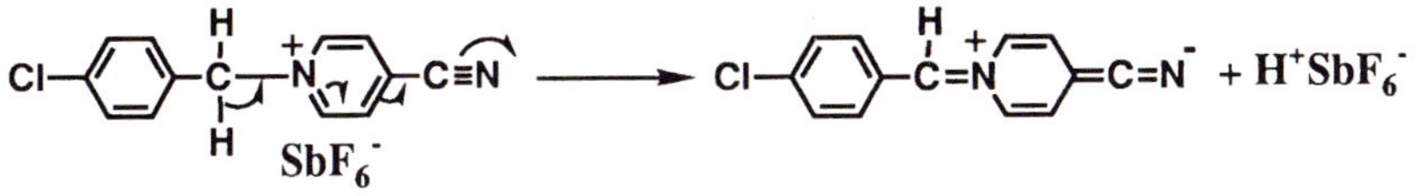

Scheme 5

initiators revealed that benzylic cation stabilization may not be the only factor determining the initiation efficiency. For instance, the activities of several *p*-substituted initiators increased in the order MeO > Cl > t-Bu > Me > H, indicating that polymer yield did not depend on the benzyl cation stability. The chlorine substituted salt **2g** is expected to exhibit much less activity since the salt possesses a *p*-electron withdrawing group which destabilizes the benzylic cation and hence lowers the benzylic cation concentration. Therefore, the unexpected activity of **2g** might show that side reactions for higher active species concentration proceeded in the cationic polymerization process using this salt. These side reactions should not include the reaction generating benzyl cations and *p*-cyanopyridines, because polymers obtained with this salt contain neither benzyl nor *p*-cyanopyridinium moieties.

One possibility for these side reactions might be to form ylide compounds, as shown in Scheme 5.

Protons generated this way may then initiate the polymerization:

$H^+ + M \longrightarrow$ **Polymer**

Interestingly, benzylanilinium salts of structure.

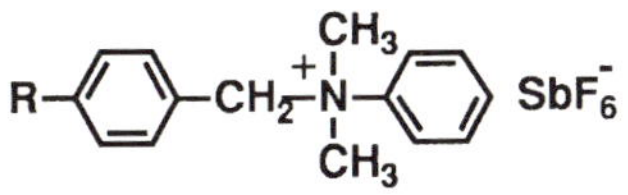

R=MeO, *t*-Bu, Me, Cl

having the same *p*-substituents showed activities in the order MeO > t-Bu > Me > Cl which was governed by benzyl cation stability. Notably, these salts will not produce ylide compounds from their structures even when they contain *p*-electron-withdrawing groups such as Cl. Consequently, the polymer yield of the salt might depend on the reactivity of the benzyl moieties.

2.2.2 Effect of Pyridine Moiety

2.2.2.1 Initiation

The polymerization of BOE with *N*-benzyl pyridinium salts with various *p*-substituents in the pyridinum moiety was studied [26] in order to observe the

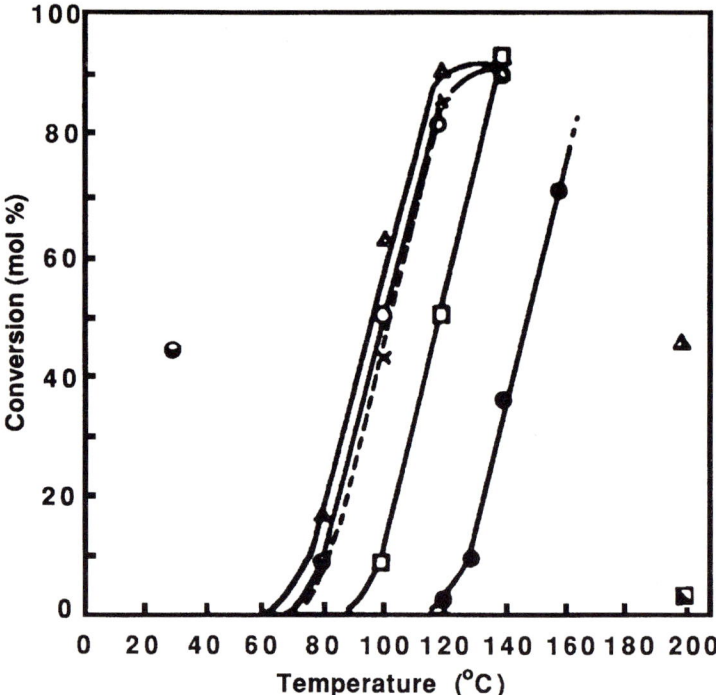

Fig. 1. Effect of temperature and initiator concentration, and substitution on the bulk polymerization of GPE with several pyridinium salts (3 mol %) for 2 h: (●) **2d**; (○) **2e**; (△) **2f**; (×) **2j**; (◓) **2k** (1 mol%); (□) **2l**; **2b** (◣); **2c** (▲)

electrostatic effect. The polymerization activity of N-benzyl p-substituted pyridinium salts decreased in the order $CN > H > CH_3 > N(CH_3)_2$, indicating that the electron withdrawing group enhanced the leaving ability of the pyridine moiety from the salt.

Figure 1 shows the effects of temperature, and concentration of the initiator, and the substitution on the polymerization of GPE. Interestingly, little or no polymerization proceeded with unsubstituted p-cyano pyridinium salt **1a** below 120 °C, whereas the rate of polymerization increased with the substituted salts.

The position of the substituents on the pyridine moiety also affects the activity of the initiator. For example, p-MeO benzyl o-CN pyridinium salt **2k** is 100 times more reactive than the corresponding p-CN derivative **2f**. Thus, introduction of the o-cyano group into pyridine nucleo causes a large enhancement of the initiator activity. The reason for the activity enhancement by this modification can be accounted for by both electronic and steric effects of the o-cyano group of the pyridine group. In a typical comparison of the initiation activities of p-CN **2f** and m-COOMe **2l** substituted p-methoxy benzyl pyridinium salts in the polymerization of GPE, it was observed that p-substituted salt **2f** is approximately 10 times more reactive than the corresponding o-substituted salt **2l**. In this case the steric effect of the substituents is negligible. A similar tendency was observed with p-CN **2d** and p-Me **2b** substituted pyridinium salts. This is in good accordance with the order of pKa of the corresponding protonated pyridines [39], i.e. that of the leaving ability in the initiation step. A different situation was encountered in the case of the salts having o- and p-Me substituents on the pyridine ring. Although both salts have similar pKa values, p-substituted salt **2b** was 15 times more reactive than the o-substituted one **2c**, indicating the importance of the steric effect [40] for the initiation in this particular case.

It was also observed that the electronic effect seems to predominate over the steric effect on the activation by changing the cyano group from the para to the ortho position.

2.2.2.2 Propagation

As discussed above, the structure of the pyridine moiety appears to affect the initiation, although the observed conversion of the monomer results from the overall polymerization. The pyridine moiety should play an important role not only in the initiating system but also in the propagation step, since it is the most basic species in a cationic polymerization system. Thus, the behavior of the pyridine moiety which was formed during the initiation step was investigated in detail [26]. Bulk and solution polymerization of GPE was strongly suppressed in the presence of additional o-cyanopyridine and p-cyanopyridine indicating interaction of pyridine moiety with propagating ends. NMR studies of the polymerization mixture in which propyleneoxide was used as a representative monomer revealed that this interaction was weaker in the case of o-cyanopyridine that of p-cyanopyridine.

According to the various results obtained a plausible mechanism of the cationic polymerization with N-benzylpyridinium salts can be postulated as depicted below (Scheme 6).

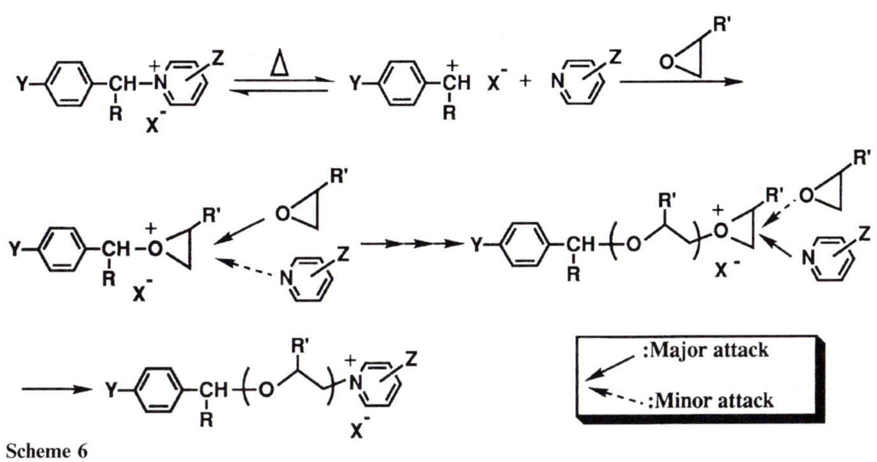

Scheme 6

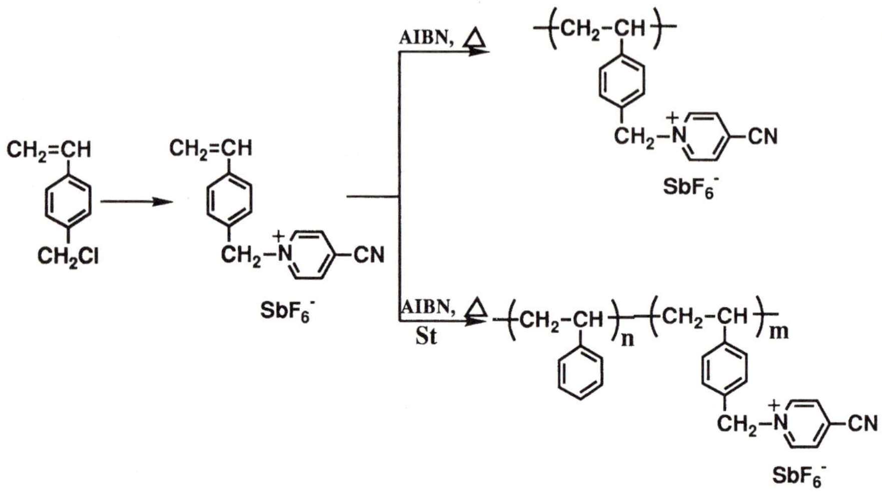

Scheme 7

2.3 Grafting onto Polystyrene Possessing N-Benzyl Pyridinium Salt Side Groups

In view of the fact that benzylic cations are responsible for the initiation, graft copolymerization is feasible when the polymer supported pyridinium salts are employed. Endo and co-workers [41] prepared vinyl monomers having pyridinium salt structure. In this case, p-chloromethyl styrene was used as the benzyl halide and the general synthetic strategy was followed. The polymerizable salt was then homo and copolymerized with styrene (Scheme 7).

The homopolymer of the pyridinium salt monomer readily initiates the cationic polymerization of BOE upon heating to around 120 °C, to yield grafted copolymers. The copolymer with styrene similarly catalyzed the polymerization to form the corresponding grafted copolymers. The initiation activity of the copolymer in the polymerization was higher than that of the homopolymer but lower than that of the low-molar mass analogue, N-benzyl-p-cyanopyridinium hexafluoroantimonate.

Another practical value of N-benzyl pyridinium salts in synthetic applications was also demonstrated by Endo and co-workers [42] who utilized these salts as catalysts for acetalization of carbonyl compounds.

3 N-Alkoxy Pyridinium Salts as Photoinitiators for Cationic Polymerization

Most industrial applications involving photopolymerization processes are based on free radical polymerization [43]. The relatively well advanced state of free radical polymerization is mainly due to the availability of a wide range of photoinitiators. There has been a growing interest in the development of initiators for corresponding cationic polymerization in recent years [44, 45]. Diaryliodonium and triarylsulphonium salts are the most important cationic photoinitiators and their photochemistry has been studied in detail [2, 6]. Recently, N-alkoxy pyridinium salts have also been shown to act as photo-initiators for cationic polymerization. In the following section, their synthesis, role in initiation from the viewpoint of wavelength selectivity and use in polymer synthesis will be discussed.

3.1 Synthesis

Reaction of pyridine-N-oxides with triethyloxonium salt in methylene chloride or chloroform gives directly N-alkoxy pyridinium salts with high yield. Quinolinium salts can also be prepared from the corresponding N-oxides [46]. The synthetic procedure does not require the anion exchange reaction since triethyl oxonium salt possesses a non-nucleophilic counter anion, i.e. PF_6^-:

$$\underset{\underset{O^-}{N^+}}{\bigcirc} + Et_3O^+PF_6^- \longrightarrow \underset{\underset{OEt}{N^+}}{\bigcirc} PF_6^- + Et_2O$$

By phenyl substitution on the pyridine moiety or by using isoquinolinium-N-oxides the absorption of the pyridinium salts can be extended to longer wavelengths [47]. Table 3 summarizes the absorption characteristics of the pyridinium salts which are used as cationic photoinitiators. In Fig. 2 the absorption maxima are compared to the emission lines of mercury. The photosensitivity of these pyridinium salts lies in the short wavelength region of the UV spectrum. Thus, pyridinium salts have to be used with sensitizers or free radical sources in order to extend their sensitivity into the region between 350–400 nm.

3.2 Initiation of Cationic Polymerization

3.2.1 Direct Initiation

Pyridinium salts act directly as photoinitiators provided the irradiation is performed at wavelengths corresponding to their absorption bands [47, 48]. A mechanism put forward to explain the ability of pyridinium salts such as p-phenyl pyridinium salt **EPP**$^+$ to induce the polymerization of appropriate monomers is shown in Scheme 8.

Upon photolysis, pyridinium salts undergo cleavage of the carbon-nitrogen bond to form pyridinium radical cation and alkoxy radical. The nitrogen centered radical cations were spectroscopically detected by laser flash photolysis

Table 3. UV absorption of N-alkoxy pyridinium salts

Denotation	Structure	Wavelength λ_{max} (nm)	Extinction coefficient (ε)
$EMP^+PF_6^-$		266	5925
$ECP^+PF_6^-$		270	–
$EIQ^+PF_6^-$		337	4218
$EPP^+PF_6^-$		310	21440

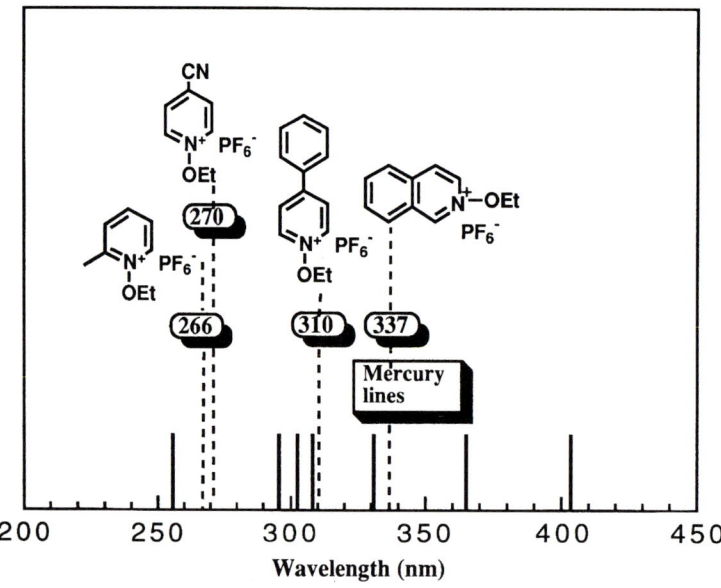

Fig. 2. Comparision of absorption maxima of N-alkoxy pyridinium salts with the emission lines of mercury

studies and were found to be highly reactive with various nucleophilic substances, including monomers like cyclohexene oxide and butyl vinylether [49]. The rate of reaction of the radical cations with nucleophiles were found to be in the range 10^6 to 10^9 $l\,mol^{-1}\,s^{-1}$. In addition to the radical cation intermediates, the

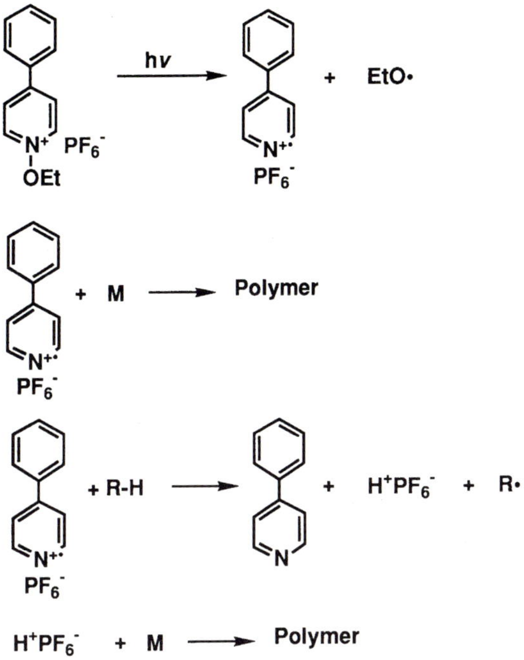

Scheme 8

Brønsted acid formed by the reactions shown in Scheme 8 may initiate the polymerization.

3.2.2 Indirect Initiation

3.2.2.1 Free Radical Promoted Cationic Polymerization

Electron donating free radicals can reduce pyridinium salts inducing their decomposition and formation of reactive carbocations [38, 50]:

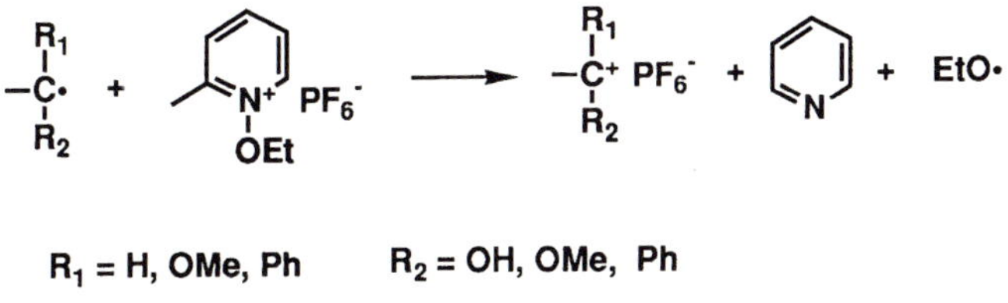

R_1 = H, OMe, Ph R_2 = OH, OMe, Ph

Electron donating radicals may be produced thermally or photochemically. Benzoin derivatives and phenylazotriphenylmethane were found to be effective to generate electron donating radicals, directly by a photochemical process (Scheme 9).

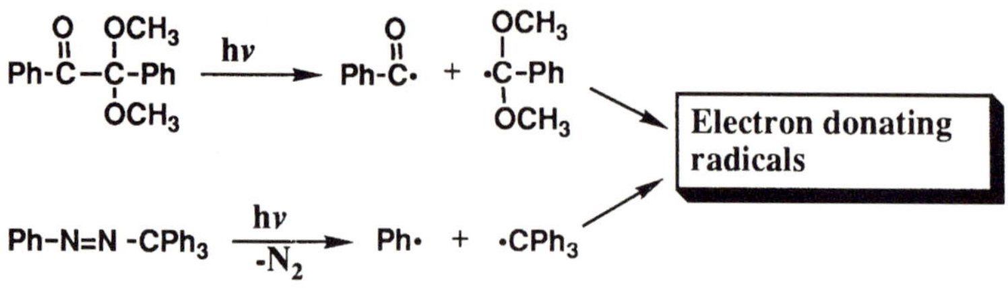

$$Ph-\underset{O}{\overset{O}{C}}-\underset{OCH_3}{\overset{OCH_3}{C}}-Ph \xrightarrow{h\nu} Ph-\overset{O}{C}\cdot + \cdot\underset{OCH_3}{\overset{OCH_3}{C}}-Ph$$

$$Ph-N=N-CPh_3 \xrightarrow[-N_2]{h\nu} Ph\cdot + \cdot CPh_3$$

→ Electron donating radicals

Scheme 9

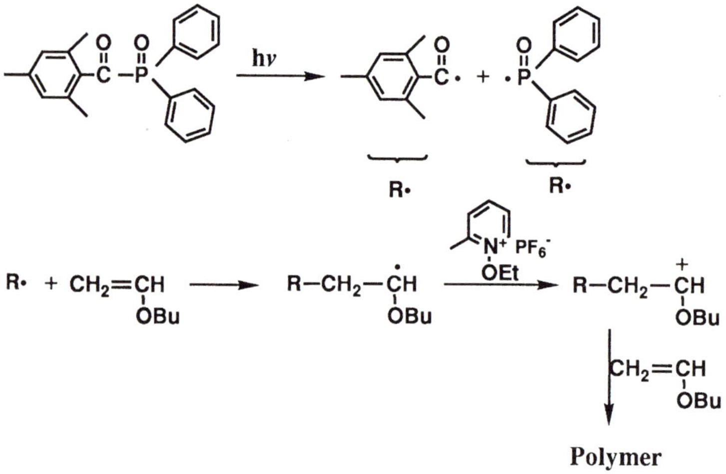

Scheme 10

Free radical promoted, cationic polymerization also occurs upon irradiation of pyridinium salts in the presence of acylphosphine oxides. But phosphonyl radicals formed are not oxidized even by much stronger oxidants such as iodonium ions as was demonstrated by laser flash photolysis studies [51, 52]. The electron donor radical generating process involves either hydrogen abstraction or the addition of phosphorus centered or benzoyl radicals to vinyl ether monomers [53]. Typical reactions for the photoinitiated cationic polymerization of butyl vinyl ether by using acylphosphine oxide-pyridinium salt combination are shown in Scheme 10.

In the case of benzophenone, electron donor radicals are formed by the hydrogen abstraction of triplet benzophenone from either monomer or deliberately added substances:

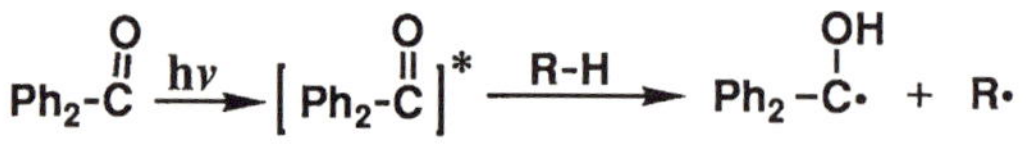

$$Ph_2-\overset{O}{C} \xrightarrow{h\nu} [Ph_2-\overset{O}{C}]^* \xrightarrow{R-H} Ph_2-\overset{OH}{C}\cdot + R\cdot$$

Ketyl radicals thus formed undergo rapid oxidation by pyridinium ions, resulting in the formation of Brønsted acid which initiates the polymerization:

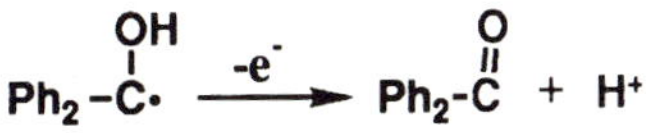

$$Ph_2\text{-}\overset{OH}{\underset{|}{C}}\cdot \xrightarrow{-e^-} Ph_2\text{-}\overset{O}{\underset{\|}{C}} + H^+$$

Notably, benzophenone is not consumed since it is regenerated in the process. o-phthalaldehyde is another interesting promoter for cationic polymerization based on pyridinium salt chemistry. Scanio and co-workers reported that UV irradiation of o-phthalaldehyde leads to the formation of 1,4-biradical transients via intramolecular hydrogen abstraction [55]:

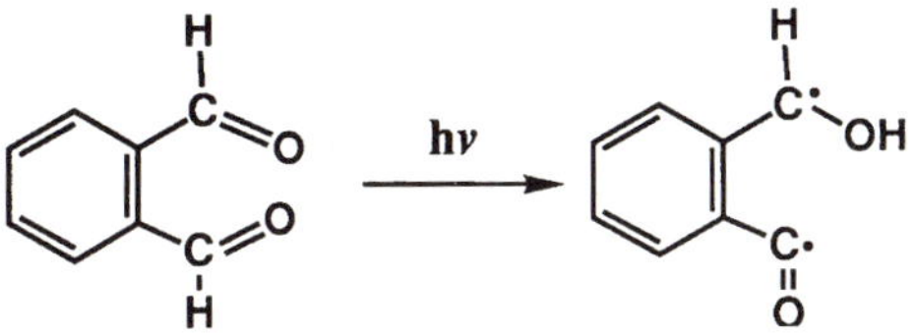

The hydroxybenzyl radical formed is a good electron donor radical and can be oxidized by suitable oxidants. The effectiveness of such a redox process was evidenced by a series of flash photolysis experiments in which the biradical was trapped by paraquat (1,1'-dimethyl-4,4'-bipyridinium) dications, PQ^{2+}, according to the following reaction:

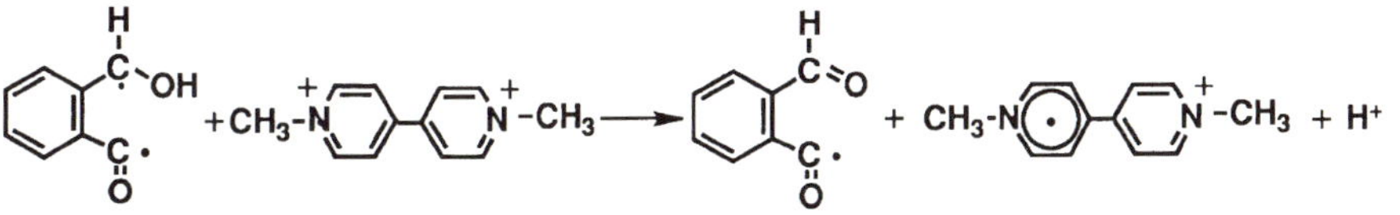

The optical absorption spectra of characteristic radical cation, $PQ^{+\cdot}$, at 630 nm was recorded [56]. It has recently been reported by Yagci and Denizligil [57] that pyridinium ions such as EMP^+ also act as a powerful oxidizing agent for the hydroxybenzyl radical to yield reactive species capable of initiating cationic polymerization (Scheme 11).

Polymerization of several representative monomers of vinyl and cyclic ethers was shown to be initiated efficiently by this particular two component system in which o-phthalaldehyde was used as promoter (Table 4). The proposed mechanism was further supported by the polymerization experiments in the presence of a strong scavenger, 2,6,6-di-t-butyl-4-methylpyridine (DBMP). Experiments in the presence of DBMP failed to produce any precipitable polymer indicating that initiation involves protons generated according to the above reaction.

It should be pointed out that o-phthalaldehyde is a suitable free radical source, without requirement of an additional hydrogen donor, for the reduction of pyridinium ions, although the initiation mechanism is quite similar to that described for benzophenone.

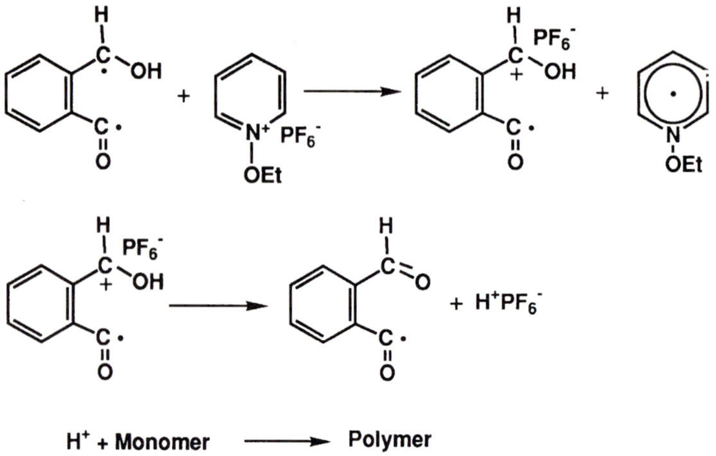

[Scheme 11 diagram]

H⁺ + Monomer ⟶ Polymer

Scheme 11

Substituted vinyl bromides in conjuction with pyridinium ions were also shown to be effective in generating initiating species [57]. The polymerizations were initiated by UV irradiation at wavelengths corresponding to the absorptions bands of the vinyl bromides where pyridinium salts absorb light only weakly or not at all. It is postulated that the photolysis of vinyl bromide yields an intermediate which rapidly reacts with pyridinium ion thus forming a species capable of initiating the cationic polymerization of oxirane compounds and vinyl ethers. According to Kitamura et al. [58] the photolysis of vinyl bromides involves, in the primary step, the homolytic cleavage of carbon-bromine bonds to form vinyl and bromine radicals. Vinyl cations formed by the subsequent electron transfer are incapable of initiating the cationic polymerization since the accompanying bromide ions are highly nucleophilic and rapidly neutralize the cations formed. Therefore, a mechanism explaining the role of the pyridinium salt should consider reactions of vinyl radicals with the pyridinium ions. Vinyl radicals are oxidized by pyridinium ions and the vinyl cations formed in this way are accompanied with the non-nucleophilic counteranion of the salt ensuring successful polymerization (Scheme 12).

An alternative mechanism might be based on electron transfer from electronically excited vinyl bromide molecules to pyridinium ions (Scheme 13).

Free radical induced cationic polymerization may also be performed by irradiating monomer solutions containing polysilane and pyridinium ions [59]. Polysilanes have strong ultraviolet absorption in the 300–350 nm region and rapid photodegradation occurs upon irradiation at this band [60]:

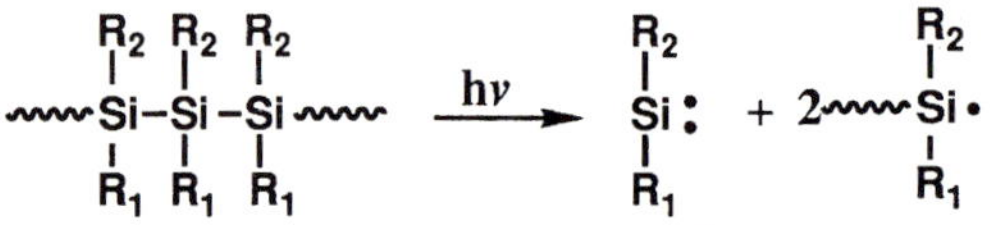

$$\sim\sim\text{Si}(R_2)(R_1)-\text{Si}(R_2)(R_1)-\text{Si}(R_2)(R_1)\sim\sim \xrightarrow{h\nu} \text{Si}(R_2)(R_1): + 2\sim\sim\text{Si}(R_2)(R_1)\cdot$$

Table 4. Photopolymerization[a] of various monomers using $EMP^+PF_6^-$ and o-phthalaldehyde, $\lambda = 350$ nm

Monomer[b]	Irradiation Time (min)	Monomer (mol/l)	$EMP^+PF_6^-$ [mol/l] × 10^2]	o-phthalaldehyde [mol/l] × 10^2]	Conversion (%)	$Mn^c \times 10^3$ (g/mol)
BVE	45	3.75	0.53	1.26	44.5	106.5
BVE	45	3.75	0.53	–	0	–
CHO	45	6.6	3.7	4.67	38.6	11.71
CHO[d]	45	6.6	3.7	4.67	0	–
CHO	45	6.6	3.7	–	0	–
NVC	10	0.97	3.2	3.7	85.2	121.4
4-VCHD	120	5.2	2.8	4.1	100	–[e]
4-VCHD	120	5.2	2.8	–	0	–

[a] In CH_2Cl_2 at room temperature
[b] BVE: butyl vinylether; CHO: cyclohexeneoxide; NVC: N-vinylcarbazol; 4-VCHD: 4-vinyl cyclohexenedioxide
[c] Determined by GPC
[d] The sample contained 2,6-di-t-butyl,4-methyl pyridine (9.1×10^{-2} mol/l)
[e] Gelled

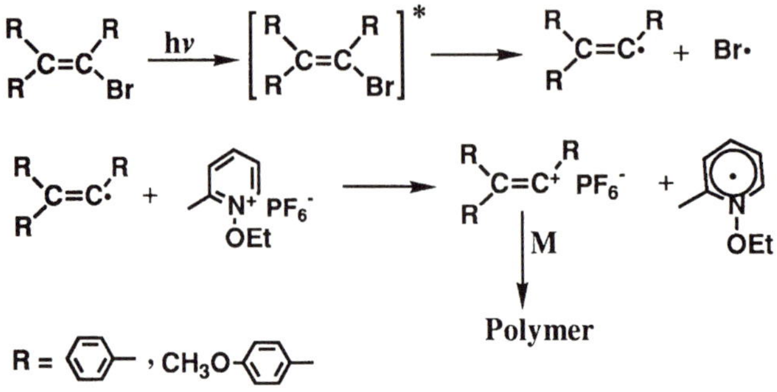

Scheme 12

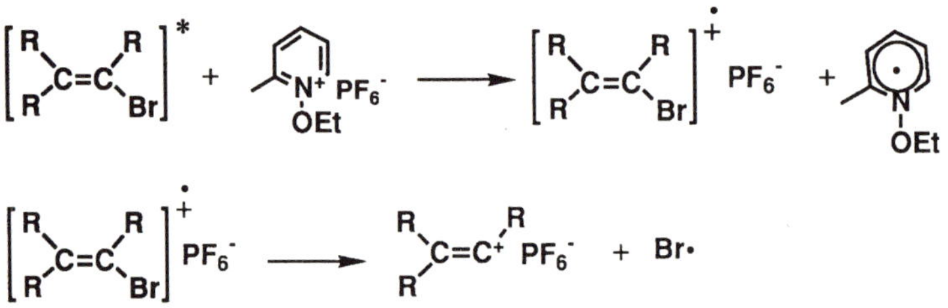

Scheme 13

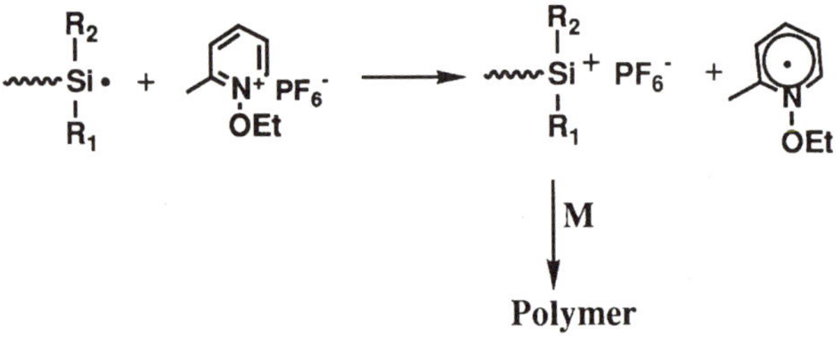

Scheme 14

When polysilanes are used in free radical polymerization, silyl-type radicals formed by chain scission are the primary initiating radicals [61, 62]. However, in the presence of pyridinium ions these radicals undergo oxidation to yield reactive cations capable of initiating the polymerization of appropriate monomers (Scheme 14).

Similar to pyridinium salts, iodonium salts can be reduced by silyl radicals generated by photolysis of polysilanes. However, due to the tail absorption bands of iodonium salts at about 300 nm, polysilanes with relatively longer

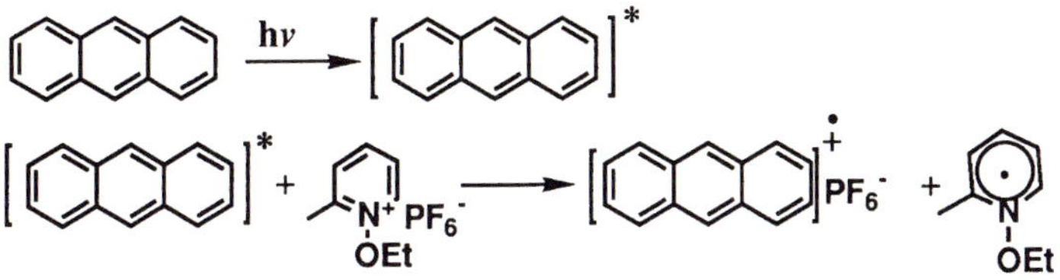

Scheme 15

wavelength absorption characteristics such as polyphenylbiphenylsilane are used [63]. As indicated earlier, the promoting radicals may also be produced thermally. Thermal decomposition of phenylazotriphenylmethane is similar to that described for the photolysis and produces triphenylmethyl radicals which can be directly converted to initiating cations. However, a different situation is encountered in the case of well known thermal free radical initiators such as 2,2′-azobisisobutyronitrile (AIBN) and dibenzoylperoxide. Upon heating these compounds generate free radicals which are not efficient reductants. The electron donor radicals required for successful oxidation are therefore produced by hydrogen abstraction or addition to vinyl ether double bond in a manner similar to that described for phosphorus centered radicals.

3.2.2.2 Photosensitized Cationic Polymerization

Various compounds were shown to sensitize the photochemical decomposition of pyridinium salts. Photolysis of pyridinium salts in the presence of sensitizers such as anthracene, perylene and phenothiazine proceeds by an electron transfer from the excited state sensitizer to the pyridinium salt. Thus, a sensitizer radical cation and pyridinyl radical are formed as shown for the case of anthracene in Scheme 15. The latter rapidly decomposes to give pyridine and an ethoxy radical. Evidence for the proposed mechanism was obtained by observation of the absorption spectra of relevant radical cations upon laser flash photolysis of methylene chloride solutions containing sensitizers and pyridinium salt [64]. Moreover, estimates of the free energy change by the Rehm-Weller equation [65] give highly favorable values for anthracene, perylene, phenothiazine and thioxanthone sensitized systems, whilst benzophenone and acetophenone seemed not to be suitable sensitizers (Table 5). The failure of the polymerization experiments sensitized by benzophenone and acetophenone in the absence of a hydrogen donor is consistent with the proposed electron transfer mechanism.

Various mechanisms for the initiation of cationic polymerization are feasible. Photosensitizer radical cations ($PS^{+\cdot}$) may react directly with the monomer M:

$$PS^{+\cdot} + M \longrightarrow {}^{\cdot}PS-M^+ \xrightarrow{nM} {}^{\cdot}PS-(M)_{\overline{n}}-M^+$$

Table 5. The free energy changes[a] (ΔG) of the electron transfer reaction of photoexcited sensitizers with pyridinium ions

Sensitizer	ΔG (kJ mol^{-1})	Photosensitization
Benzophenone	+ 39.8	No
Acetophenone	+ 41.2	No
Thioxanthone	− 44.2	Yes
Anthracene	− 144.4	Yes
Perylene	− 121.8	Yes
Phenothiazine	− 112.9	Yes

[a] Estimated by using $\Delta G = f_c(E^{ox}_{1/2} - E^{red}_{1/2}) - E(PS)^*$ [65]; $E^{ox}_{1/2}$ = halfwave oxidation potential of sensitizer, $E^{red}_{1/2}$ = halfwave reduction potential of EMP$^+$, E(PS*) = excitation energy of sensitizer, f_c = the conversion factor

Alternatively, PS$^{+\cdot}$ may abstract hydrogen from the solvent or the monomer and the resulting intermediate dissociates:

$$\overset{\cdot}{PS^+} + R\text{-}H \longrightarrow H\text{-}PS^+ + R\cdot$$

$$H\text{-}PS^+ \longrightarrow PS + H^+$$

Protons generated in this way add to the monomer, thus forming species capable of initiating the cationic polymerization of M:

$$H^+ + M \longrightarrow H^+\text{-}M^+ \xrightarrow{nM} H\text{-}(M)_{\overline{n}}M^+$$

Support for direct initiation by PS$^{+\cdot}$ was recently obtained by experiments on the polymerization of cyclohexene oxide using anthracene labelled polytetrahydofuran as the sensitizer. In this case poly(tetrahydrofuran-b-cyclohexeneoxide) is formed [66].

For the thioxanthone sensitization, depending on the monomer and pyridinium ion concentration, both electron transfer and hydrogen abstraction mechanisms are proposed [66].

3.2.2.3 Photoinitiation by Charge Transfer Complexes

It is well known that cyano derivatives of anthracene form charge transfer (CT) complexes with certain aromatic compounds. It was reported [67] that the radical cations formed upon irradiation of these complexes played an important role in initiation of cationic polymerization of cyclic ethers. Pyridinium salts were also found [68] to form CT complexes with hexamethyl benzene and trimethoxy benzene which result in the formation of a new absorption band at longer wavelengths where both donor and acceptor molecules have no absorption. This way the light sensitivity of the pyridinium salts may be extended towards the visible range. According to the results obtained from the

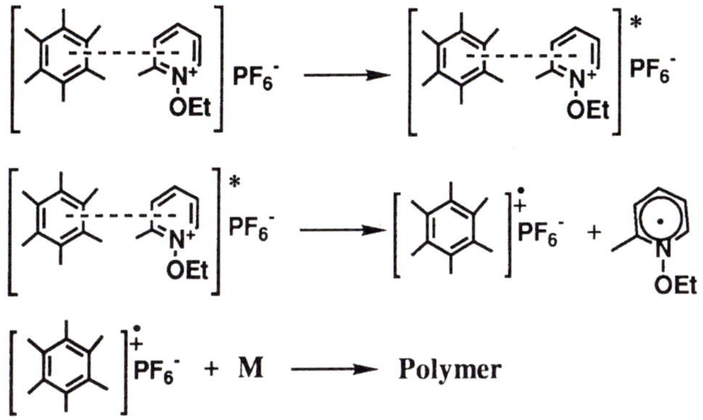

Scheme 16

polymerization experiments in the presence and absence of a proton scavenger, the following mechanism for the initiation was proposed (Scheme 16).

Cyclic monomers such as cyclohexene oxide were readily polymerized upon irradiation of the CT complexes of pyridinium salts whereas spontaneous polymerizations were observed upon mixing with strong electron donating monomers such as butyl vinylether and N-vinyl carbazole. These monomers are known to form CT complexes themselves with electron acceptors which may interfere with the rapid polymerization observed.

3.2.2.4 Redox-Initiated Cationic Polymerization

Ascorbic acid (AH_2) and its 5- and 6-acyl derivatives are known to be suitable reducing agents. The reducing properties of AH_2 is due to the enediol structure as shown below:

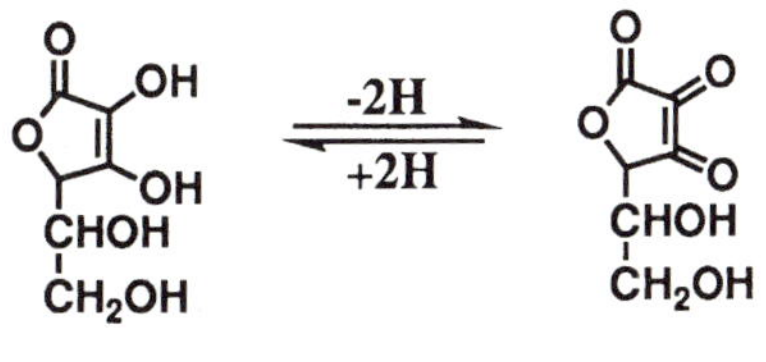

AH_2 DA

Crivello and Lam [69] have reported that the diaryliodonium salt-ascorbate redox system readily initiates the cationic polymerization of appropriate monomers. N-Alkoxy pyridinium salts were also shown [70] to participate in this redox process. The polymerization mechanism depicted below is quite similar to that described for the iodonium salts (Scheme 17).

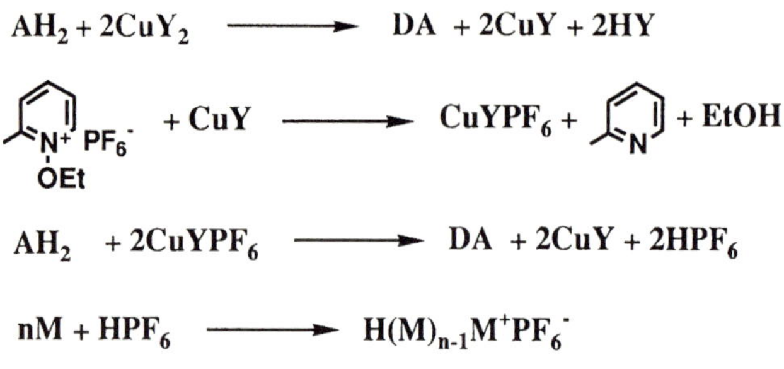

$$AH_2 + 2CuY_2 \longrightarrow DA + 2CuY + 2HY$$

$$\underset{\underset{OEt}{|}}{\overset{}{\underset{N^+}{\bigcirc}}} PF_6^- + CuY \longrightarrow CuYPF_6 + \underset{N}{\bigcirc} + EtOH$$

$$AH_2 + 2CuYPF_6 \longrightarrow DA + 2CuY + 2HPF_6$$

$$nM + HPF_6 \longrightarrow H(M)_{n-1}M^+PF_6^-$$

Scheme 17

The first step in the mechanism involves the reduction of Cu(II) to Cu(I) by ascorbyl-6-hexadecanoate giving dehydroascorbic acid and a weak acid HY (benzoic acid). In fact this stage of the process has no importance since Cu(I) benzoate may directly be used to initiate the polymerization by reducing the pyridinium salt. The strong Brønsted acid formed attacks the monomer and initiates the polymerization. Notably, lower polymer yields were obtained by using pyridium salt rather than iodonium salt.

3.3 Photoinitiated Free Radical Polymerization by Using Pyridinium Salts

The use of *N*-alkoxy pyridinium salts is not limited to cationic polymerization. Since, in addition to cationic species, ethoxy radicals are also formed upon direct and sensitized irradiation of pyridinium salts (see above), pyridinium salt based photoinitiating systems may be used to initiate the polymerization of vinyl monomers that are prone to free radical polymerization. Kayaman et al. [71] recently polymerized mono- and bi-functional acrylate monomers by photosensitization of pyridinium salts. It therefore appears that pyridinium salts can promote both cationic and free radical polymerization and are, thus, eminently suitable for use in hybrid systems.

3.4 Synthesis of Block Copolymers by Using N-Alkoxy Pyridinium Salts

Polymers containing benzoin terminal groups can act as photochemical macroinitiators and are effective in photogenerating polymeric electron donor radicals. The initiation of polymerization by means of azo-benzoin initiators yields polymers with one or two benzoin end-groups according to the termination mode of the particular monomer involved [72–74]. The general synthetic procedure is depicted below as illustrated for the case styrene polymerization (Scheme 18).

$$\underset{\text{OCH}_3}{\overset{\text{O Ph}}{\underset{|}{\overset{||}{\text{PhC}}-\overset{|}{\text{C}}}}}\text{CH}_2\text{OOCCH}_2\text{CH}_2\underset{\text{CN}}{\overset{\text{CH}_3}{\underset{|}{\overset{|}{\text{C}}}}}-\text{N=N}-\underset{\text{CN}}{\overset{\text{CH}_3}{\underset{|}{\overset{|}{\text{C}}}}}\text{CH}_2\text{CH}_2\text{COOCH}_2\underset{\text{OCH}_3}{\overset{\text{Ph O}}{\underset{|}{\overset{||}{\text{C}}}}}-\text{CPh} \xrightarrow[-\text{N}_2]{\Delta} 2\, \underset{\text{OCH}_3}{\overset{\text{O Ph}}{\underset{|}{\overset{||}{\text{PhC}}-\overset{|}{\text{C}}}}}\text{CH}_2\text{OOCCH}_2\text{CH}_2\underset{\text{CN}}{\overset{\text{CH}_3}{\underset{|}{\overset{|}{\text{C}}}}}\bullet$$

$$\xrightarrow{n\,\text{St}} \underset{\text{OCH}_3}{\overset{\text{O Ph}}{\underset{|}{\overset{||}{\text{PhC}}-\overset{|}{\text{C}}}}}\text{CH}_2\text{OOCCH}_2\text{CH}_2\underset{\text{CN}}{\overset{\text{CH}_3}{\underset{|}{\overset{|}{\text{C}}}}}-\underset{\text{Ph}}{[\text{CH}_2-\text{CH}]_n}-\underset{\text{CN}}{\overset{\text{CH}_3}{\underset{|}{\overset{|}{\text{C}}}}}\text{CH}_2\text{CH}_2\text{COOCH}_2\underset{\text{OCH}_3}{\overset{\text{Ph O}}{\underset{|}{\overset{||}{\text{C}}}}}-\text{CPh}$$

Scheme 18

UV irradiation of the resulting prepolymers caused α-scission, and benzoyl and polymer bound electron donating radicals are formed in the same manner as described for the low-molar mass analogues. Electron donating polymeric radicals thus formed may conveniently be oxidized to polymeric carbocations to promote cationic polymerization of cyclic ethers. It was demonstrated that irradiation of benzoin terminated polymers in conjuction with pyridinium salts as oxidants in the presence of cyclohexene oxide makes it possible to synthesize block copolymers of monomers with different chemical natures [75] (Scheme 19).

Similarly, side chain benzoin-containing polymers were also used as promoters to yield graft copolymers [76].

Free radical promoted cationic polymerization was successfully employed [77] for the preparation of new classes of liquid crystalline (LC) block copolymers comprising a semicrystalline block, poly(cyclohexene oxide), and LC block of different structures:

block A   **block B**

$+O-\text{C}_6\text{H}_{10}+_x$ .... $+\text{CH-CH}_2+_y$
                                      |
                                      C=O
                                      |
                                      O(CH$_2$)$_n$-⟨⟩-⟨⟩-OR

BC1: n=11; R= (S)-OCCH*(Cl)i.C$_3$H$_7$
BC2: n=10; R= (S)-CH$_2$CH*(CH$_3$)C$_2$H$_5$

The synthetic scheme described previously for the preparation of block copolymers of styrene and cyclohexene oxide by employing pyridinium salts was adapted. Thus, a macroinitiator is obtained in the first step by promoted cationic polymerization. Upon heating, this macroinitiator generates macroradicals that then polymerize a mesogenic acrylate monomer by a free-radical mechanism leading to block copolymers. The resulting block copolymers possess ABA triblock structure, the cyclohexene oxide block being the A component and liquid crstalline block the B component, since the free radical

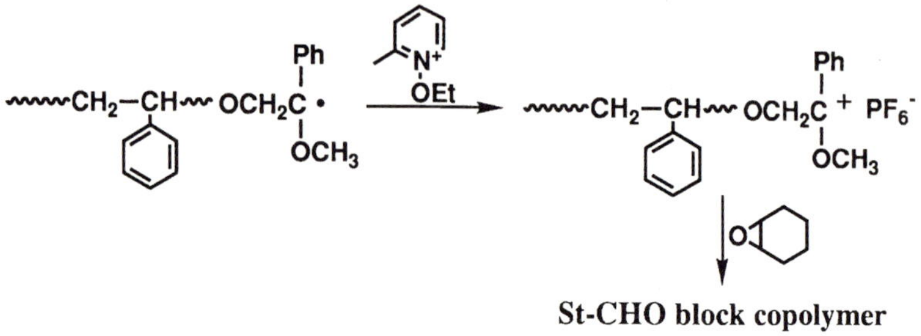

Scheme 19

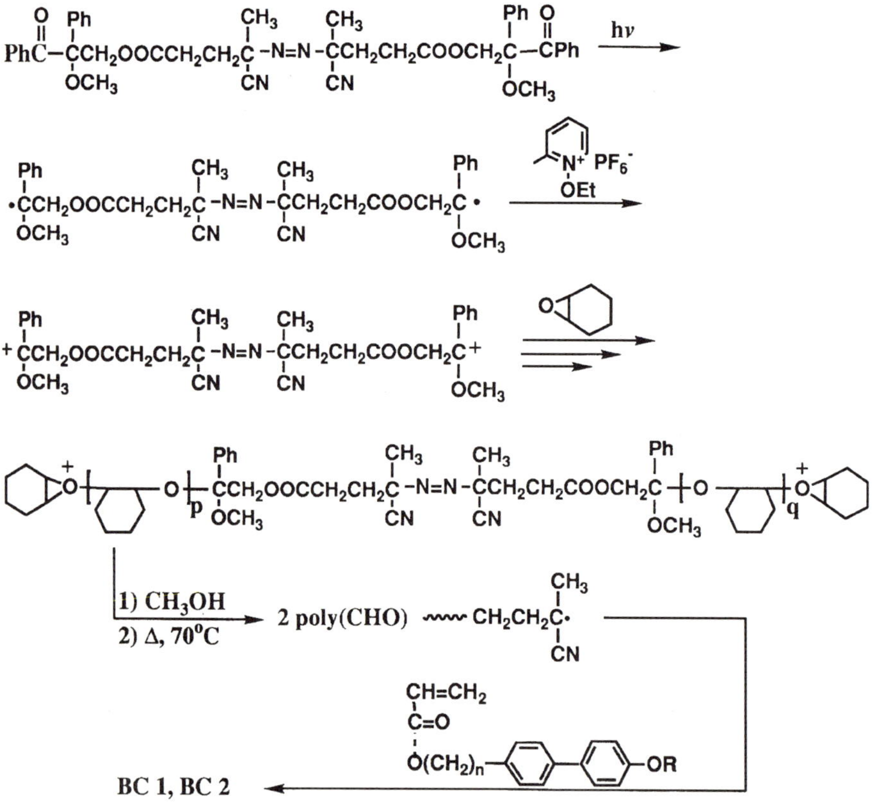

Scheme 20

polymerization of acrylate monomers terminates almost quantitatively by a combination mechanism (Scheme 20).

The block copolymers obtained are essentially microphase separated systems and form smectic mesophases, analogous to the corresponding liquid crystalline homopolymers.

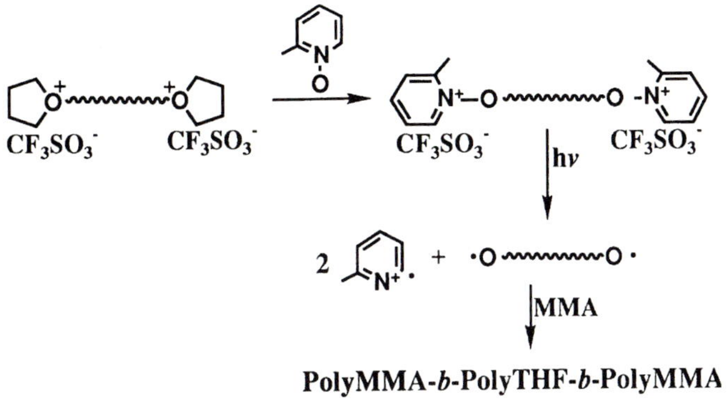

Scheme 21

PolyMMA-*b*-PolyTHF-*b*-PolyMMA

Another synthetic approach based on pyridium salt photochemistry involves the use of alkoxy radicals which are formed in both direct and sensitized decomposition of pyridinium ions in free radical polymerization [78]. Obviously, polytetrahydrofuran (PTHF), terminated by *N*-alkoxy pyridinium ions, can act as macrophotoinitiator for the polymerization of monomers such as methyl methacrylate (MMA) that readily polymerize by a free-radical mechanism. PTHF macrophotoinitiators were prepared by termination of living polymerization of THF by the corresponding *N*-oxides, The well-defined macrophotoinitiators with exact functionalities, confirmed by ^{1}H-NMR, UV-visible and g.p.c. analysis, were obtained. Upon irradiation of macroinitiators at suitable wavelengths, polymeric alkoxy radicals are produced. The overall process is shown for the pyridinium macrophotoinitiator in the following Scheme 21.

Photolysis in the presence of MMA gives quantitative yields of PMMA-*b*-PTHF-*b*-PMMA block copolymers. Block copolymers with various segment lengths were obtained by precipitating at room temperature and $-20\,°C$. Indirect photolysis using anthracene sensitizer in the presence of MMA gives rise to the formation of the same block copolymers. Again the reaction products were almost pure block copolymers.

Hizal et al. [79] have recently shown an interesting variation of pyridinium salt photodecomposition in polymer synthesis. In earlier investigations it became evident that ethanol is formed by the hydrogen abstraction of primary ethoxy radical if the photolysis is carried out in strong hydrogen donor solvents such as THF:

Hydroxy-functional telechelic polymers were readily prepared by taking advantage of the above reaction. Hydroxy end-groups could be effectively introduced into the polymers when pyridinium ion terminated PTHF irradiated in THF:

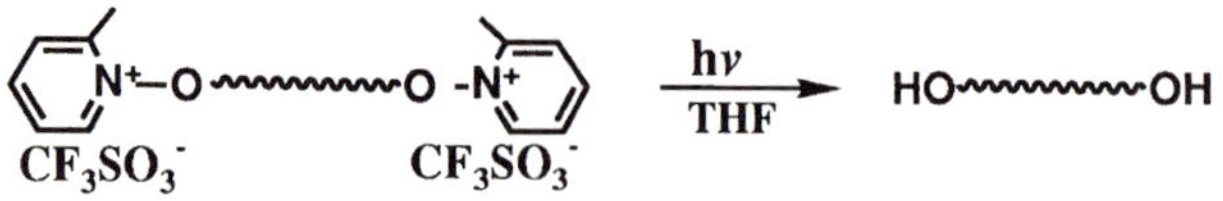

$$\text{Py}^+\text{-O}\sim\sim\sim\text{O-Py}^+ \;\; 2\,CF_3SO_3^- \xrightarrow[THF]{h\nu} HO\sim\sim\sim OH$$

Moreover, in situ polyurethane formation was performed by irradiation of the polymeric pyridinium salt in THF containing toluene diisocyanate and catalyst. It is clear that alkoxy pyridinium terminated polymers are useful materials as precursors for block copolymers and hydroxy functional telechelics. The latter are particularly attractive in photoinduced polycondensation and in applications where hydroxyl groups are needed to be protected.

4 Conclusions

N-Benzyl and N-alkoxy pyridinium salts are suitable thermal and photochemical initiators for cationic polymerization, respectively. Attractive features of these salts are the concept of latency, easy synthetic procedures, their chemical stability and ease of handling owing to their low hygroscopicity. Besides their use as initiators, the applications of these salts in polymer synthesis are of interest. As shown in this article, a wide range of block and graft copolymer built from monomers with different chemical natures are accessible through their latency.

Acknowledgement. One of the authors (Y.Y) would like thank Japan Society for Promotion of Science for supporting his stay in Tokyo Institute of Technology, Tokyo, Japan during which this article was written

5 References

1. Kennedy JP (1982) Carbocationic polymerization. Wiley Interscience, New York
2. Crivello JV (1991) In: Dietliker K (ed) Chemistry & technology of UV & EB formulation for coatings, inks & paints. SITA Technology, London, Chap III
3. Crivello JV, Lam JHW (1980) J Polym Sci Polym Chem Ed 18: 2677
4. Crivello JV, Lam JHW (1980) J Polym Sci Polym Chem Ed 18: 2967
5. Crivello JV, Lam JHW (1979) Macromolecules 10: 1307
6. Crivello JV (1984) Adv Polym Sci 62, 1
7. Endo T, Uno H (1985) J Polym Sci Polym Lett Ed 23, 359
8. Kikkawa A, Takata T, Endo T (1991) Makromol Chem 192, 655
9. Hamazu F, Akashi, Koizumi, Takata T, Endo T (1991) J Polym Sci Polym Chem 29, 1675
10. Hamazu F, Akashi, Koizumi, Takata T, Endo T (1991) J Polym Sci Polym Chem 29, 1845

11. Park J, Kihara N, Ikeda T, Endo T (1993) J Polym Sci Polym Chem Ed 31, 1083
12. Hamazu F, Akashi, Koizumi, Takata T, Endo T (1991) J Polym Sci Polym Chem 31, 1023
13. Hamazu F, Akashi, Koizumi, Takata T, Endo T (1992) J. Photopolym Sci Techn 5, 247
14. Hamazu F, Akashi, Koizumi, Takata T, Endo T (1992) Makromol Chem Rapid Commun 13, 203
15. Morio K, Murase H, Tsuciya H, Endo T (1986) J Appl Polym Sci 32, 5727
16. Lin MP, Ikeda T, Endo T (1992) J Polym Sci Polym Chem 30, 2576
17. Lin MP, Hayashi Y, Ikeda T, Endo T (1992) J Polym Mat Sci 27, 2992
18. Yagci Y, Denizligil S, McArdle C (1995) Polymer, 36, 3093
19. Schelesinger SI (1974) Photogr Sci Eng 18, 387
20. Crivello JV (1989) Radiation Curing Workshop, ACS Meeting Dallas April
21. Takata T, Takuma K, Endo T (1993) Makromol Chem, Rapid Commun 14, 203
22. Takuma K, Takata T, Endo T (1993) J Photopolym Sci Techn 6, 67
23. Takuma K, Takata T, Endo T (1993) Macromolecules 26, 862
24. Uno H, Endo T (1988) J Polym Sci Polym Lett Ed 26, 453
25. Lee SB, Takata T, Endo T (1989) Chem Let 1861
26. Lee SB, Takata T, Endo T (1991) Macromolecules 24, 2689
27. Nakano S, Endo T (1993) Prog Org Coatings 22, 862
28. Uno H, Takata T, Endo T (1988) Chem Let 935
29. Lee SB, Takata, Endo T (1990) Macromolecules 23, 431
30. Nakano S, Endo T (1995) J Polym Sci Polym Chem Ed 33, 505
31. Yagci Y, Schnabel W (1994) Macromol Symp 84, 115
32. Roloff A, Meir K, Riediker (1986) Pure Appl Chem 58, 1267
33. Meir K, Zwifel H (1986) J Radiat Curing 13, 26
34. Berner G, Kirchmayer R, Rist G, Rutsch W (1986) J Radiat Curing 13(4), 10
35. Fouassiser JP, Burr D (1990) Macromolecules 23, 3615
36. Hayase S, Onishi Y, Suzuki S, Wada M (1986) Macromolecules 19, 968
37. Abdoul Rasoul FM, Ledwith A, Yagci Y (1978) Polymer 19, 1219
38. Bottcher A, Hasebe K, Hizal G, Yagci Y, Stelberg P, Schnabel W (1991) Polymer 32, 2289
39. Lange NA (1985) In: Dean JA (ed.) Lange's Handbook of Chemistry, 13th ed., McGraw Hill, New York, pp 5–60
40. March J (1985) In: Advanced Organic Chemistry, 3rd ed.; Wiley-Interscience, New York, p 248
41. Uno H, Takata T, Endo T (1989) Macromolecules 22, 2502
42. Lee SB, Lee SD, Takata T, Endo T (1991) Synthesis 5, 368
43. Gruber HF (1992) Prog Polym Sci 17, 953
44. Pappas SP (1985) UV Curing Science and Technology, Vol. II, Technology Marketing Corp., Norwalk, CT
45. Allen NS (1989) Photopolymerizaton and Photoimaging Science and Technology, Elsevier, New York, pp 55–75
46. Reichard C (1966) Chem Ber 99, 1769
47. Yagci Y, Kornowski A, Schnabel W (1992) J Polym Sci Polym Chem Ed 30, 1987
48. Yagci Y, Kornowski A, Masonne K, Schanabel W (1991) Pending Patent Application, DE-P 4103906.8, Feb.
49. Yagci Y, Schnabel W (1993) Macromol Reports, A30 (Suppl. 3 & 4) 175
50. Bottcher A, Schanabel W, Yagci Y (1990) Pending Patent Application DE-P 4003925, Feb.
51. Yagci Y, Schnabel W (1987) Makromol Chem, Rapid Commun 8, 209
52. Yagci Y, Schnabel W (1988) Makromol Chem, Macromol Symp 13–14, 161
53. Yagci Y, Borberly B, Schnabel W (1989) Eur Polym J 25, 129
54. Scanio JC, Encinas MV, George MV (1974) J Chem Soc Perkin Trans. 2, 1, 724
55. Hyde P, Ledwith A (1974) J Chem Soc Perkin Trans, 1768
56. Yagci Y, Denizligil S (1995) J Polym Sci, Polym Chem Ed 33, 1461
57. Johnen N, Kobayashi S, Yagci Y, Schnabel W (1993) Polym Bull 30, 279
58. Kitamura T, Koboyashi S, Tanuguchi H (1986) J Am Chem Soc 108, 2641
59. Yagci Y, Kminek I, Schnabel W (1992) Eur Polym J 28, 387
60. Michl J, Downing JW, Karatsu T, McKinley AJ, Poggy G, Wallraff GM, Sooriyakumaran R, Miller RD (1988) Pure & Appl Chem 60, 959
61. West R, Wolf AR, Peterson PJ (1986) J Radiat Curing 13, 35
62. Kminek I, Yagci Y, Schnabel W (1992) Polym Bull 29, 277
63. Yagci Y, Kminek I, Schnabel W (1993) Polymer 34, 426

64. Yagci Y, Lukac I, Schnabel W (1993) Polymer 34, 1130
65. Rehm D, Weller A (1969) Bunsenges Phys Chem 73, 834
66. Dossow D, Hizal G, Zhu QQ, Yagci Y, Schnabel W (1995) Polymer submitted
67. Ohtsuka T, Yamamoto, Hayashi K (1988) J Polym Sci Polym Lett Ed, 26, 481
68. Hizal G, Yagci Y, Schnabel W (1994) Polymer 35, 2428
69. Crivello JV, Lam JHW (1981) J Polym Sci Polym Chem Ed 19, 539
70. Onen A, Yagci Y (1995) in preparation
71. Kayaman N, Onen A, Yagci Y, Schnabel W (1994) Polym Bull 32, 589
72. Onen A, Yagci Y (1990) J Macromol Sci Chem A27 743
73. Onen A, Yagci Y (1990) Angew Makromol Chem 181, 191
74. Yagci Y, Onen A (1991) J Macromol Sci Chem A28, 129
75. Yagci Y, Onen A, Schnabel W (1991) Macromolecules, 24, 4620
76. Onen A, Yagci Y (1992) Eur Polym J, 28, 721
77. Serhatli IE, Galli G, Chiellini E, Yagci Y (1995) Polym Bull, 34, 539
78. Hizal G, Yagci Y, Schnabel W (1994) Polymer 35, 4443
79. Hizal G, Sarman A, Yagci Y (1995) Polym Bull 35, 567

Editor: Prof. A. Ledwith
Received: February 1996

Synthesis of Block Copolymers by Radical Polymerization and Telomerization

B. Améduri, B. Boutevin and Ph. Gramain
Ecole Nationale Supérieure de Chimie de Montpellier, URA 1193 CNRS 8 rue Ecole Normale, 34053 Montpellier Cedex/France

The synthesis of block copolymers, cooligomers or cotelomers performed from radical polymerization and radical telomerization is discussed. In the first part, the two main methods proposed for obtaining block copolymers by radical polymerization are reviewed. The first method concerns the use of multifunctional micro- or macroinitiators that decompose according to two processes or at two different temperatures, allowing the stepwise insertion of different base units. The second method, more recently proposed, tends to control the reactivity of the growing radicals by the use of either additional stable free radicals or organometallic compounds acting as counter radical, allowing reversible termination steps. These methods still in full development and often referred to as "living" processes, allow well defined block copolymers with narrow molecular weight distributions to be produced. The second part of the review deals with methods of preparation by radical telomerization. Besides the traditional process involving the coupling of ω-functional oligomers or telomers and the bistelomerization, the dormant or living telomerization methods are also described. In this last case, the "living" character is achieved by the use of well-chosen transfer agents with specific cleavable bonds leading to reversibility of the termination step.

List of Symbols and Abbreviations		88
1	**Introduction**	90
2	**Synthesis of Block Copolymers by Radical Polymerization**	91
	2.1 Multifunctional Free Radical Initiators	91
	2.1.1 Micromolecular Initiators	91
	2.1.2 Polymeric Initiators	95
	2.2 Controlled Radical Polymerization Involving Stable Free Radicals	98
	2.2.1 Nitroxyls and related Compounds	99
	2.2.2 Organometallic complexes	103
	2.3 Conclusion	105
3	**Synthesis of Block Copolymers by Radical Telomerization**	105
	3.1 Uncontrolled Radical Telomerization	106
	3.1.1 Bistelomerization	106
	3.1.2 Coupling of Monofunctional Telomers	110
	3.1.3 Conclusion	112

3.2 Controlled Radical Telomerization 113
    3.2.1 C–Halogen Cleavage 113
    3.2.2 C–C Cleavage 119
    3.2.3 C–S Cleavage 123
    3.2.4 S–S Cleavage 125
    3.2.5 Si–Si Cleavage 134
    3.2.6 Conclusion 135

4 Conclusion 135

5 References 137

List of Symbols and Abbreviations

AA	acrylamide
acac	acetylacetonate
alt	alternating
AIBN	azobisisobutyronitrile
Ar	aromatic group
Bd	butadiene
Bu	n-butyl
BuA	n-butyl acrylate
BuMA	n-butyl methacrylate
BzMA	benzyl methacrylate
CMSty	chloromethyl styrene
CR	counter-radical
C_T	transfer constant
CTFE	chlorotrifluoroethylene
DBP	dibenzoyl peroxide
DPn	average degree of polymerization in number
EA	ethyl acrylate
HEA	2-hydroxyethyl acrylate
HEMA	2-hydroxyethyl methacrylate
HFP	hexafluoropropene

iBu	isobutyl
Ip	isoprene
ITP	iodine transfer polymerization
k	rate constant
k_i	rate constant of initiation
M	monomer
[M]	concentration of monomer
MA	methyl acrylate
MAO	methyl aluminoxane
M_i	monomer i
MMA	methyl methacrylate
Mn	average molecular weight in number
Mw	average molecular weight in weight
NMR	nuclear magnetic resonance
PBd	poly(butadiene)
PBuA	poly(butyl acrylate)
PCL	poly(ε-caprolactone)
PDMS	poly(dimethylsiloxane)
PEA	poly(ethyl acrylate)
PEG	poly(ethylene glycol)
PEO	poly(ethylene oxide)
PFA	poly(fluoroacrylate)
PLLA	poly(L-lactide)
PMA	poly(methyl acrylate)
PMMA	poly(methyl methacrylate)
PPO	poly(propylene oxide)
PS	poly(styrene)
PVA	poly(vinyl alcohol)
PVAc	poly(vinyl acetate)
R·	alkyl radical
R	alkyl group
RT	room temperature
R_FI	perfluoroalkyl iodide
Sty	styrene
tBu	*tert*-butyl
Tempo	2,2,6,6-tetramethyl piperidinyl-l-oxyl
TFE	tetrafluoroethylene
TD	thiuram disulfide
THF	tetrahydrofuran
Tg	glass transition temperature
TX	xanthogen disulfide
UV	ultra violet
VAc	vinyl acetate
VDF	vinylidene fluoride
Δ	heating

1 Introduction

Progress in polymer science and technology is increasingly dependent on the availability of materials to meet a required specific performance. Actually, the scientist has to fulfill requirements in terms of processability, durability, resistance to environment, cost-effectiveness, and mechanical performance to find new materials for specific applications. In this way, macromolecular engineering has become a useful tool to design well controlled architectured polymers: telechelics, cycles, networks, block and grafted copolymers, star shape polymers and dendritic structures. Among them, block copolymers have received much attention as "novel polymeric materials" with multi-components [1–5] since they are made of different polymeric sequences linked together. The reason for their importance comes from their unique chemical structure that brings new physical and thermodynamical properties related to their solid-state and solution morphologies. Frequently. block copolymers exhibit phase separation producing a dispersed phase consisting of one block type in a continuous matrix of the second block type [6–8]. Their unusual colloidal and mechanical properties allow modification of solution viscosity, surface activity or elasticity and impact resistance. Thus, several block copolymers have produced a wide range of materials with tailorable properties depending on the nature and length of homosequences. They have found significant commercial applications [3–6, 8–14] such as adhesives and sealants, surface modifiers for fillers and fibers, crosslinking agents for elastomers, additives for resin gelification and hardening, compatibilizing agents or stable emulsions of homopolymer blends that can find applications in recovering and recycling plastic waste.

A number of synthetic methods have been successfully developed for the synthesis of block copolymers. They include polycondensation, anionic, cationic, coordinative and free-radical polymerizations and also mechanochemical synthesis. Despite the exceptional amount of attention paid to the prospects of various catalytic systems, radical polymerization has not lost any of its importance, particularly in this area. Its competitiveness with other methods of conducting polymerization are attributable to the simplicity of the mechanism and good reproducibility. Actually, the extensive use of free radical polymerization in practice is well understood when considering the ease of the process, the soft processable conditions of vacuum and temperature, the fact that reactants do not need to be highly pure and the absence of residual catalyst in the final product. Thus, it can be easily understood that more than 50% of all plastics have been produced industrially via radical polymerization.

Methods of obtaining block copolymers by radical processes have been developed rather lately about other processes, and especially ionic methods. This may be due to the nature of the radical, which is an intermediate with a very short lifetime, and a very high non-selective reactivity. These characteristics do not favor a well-controlled architecture as in the case of living carbanions appearing in anionic polymerization. However, the recent development of new methods to control the reactivity of radicals and to give a living character to the growing macroradicals offers new fascinating possibilities.

The objective of this review concerns strategies of synthesis of block copolymers through free radical methods including their advantages and drawbacks. First, various methods involving radical polymerization are described. They concern either the use of multifunctional organic micro- or macromolecular initiators which decompose in a two-step process in order to initiate the polymerization of two different monomers, or more recent systems using counter radicals or organometallic compounds able to stabilize the growing radicals reversibly, insuring a pseudo-living character and a better control of the polymerization. The second part is devoted to the radical telomerization: the traditional one involves the coupling of monofunctional telomers and the bistelomerization, whereas the pseudo-living telomerization shows how adequate telogens and catalysts are able to provide an accurate control of such a process.

2 Synthesis of Block Copolymers by Radical Polymerization

In radical polymerization, general methods to obtain block copolymers are based on an initiation step and a propagation step. In the first case, a macromolecular initiator is first synthesized either by a radical method or by step polymerization, then this macromolecular initiator initiates the copolymerization of a second monomer. As will be seen, this method allows the synthesis of a large variety of interesting structures but leads most often to mixtures of di-, tri- or multiblock structures due to the uncontrolled termination step and these are difficult to separate. In the second method, block structures are obtained by varying the propagation step. This requires rigorous control of this step which can only be obtained by strict control of the reactivity of the growing macroradicals. The recent progresses reached in this domain open new possibilities.

2.1 Multifunctional Free Radical Initiators

2.1.1 Micromolecular Initiators

A multifunctional initiator includes in its chemical structure too or more chemical functions able to generate active radicals. By choosing properly the nature of these functions, it is possible to produce initiating radicals in two or more steps either thermally, photochemically or in the presence of electronic radiation. Various multifunctional initiators have been synthesized and successfully used. Work in this field has been very well summarized by Ivanchev [15] and Simionescu et al. [13]. The initiators have been classified according to their ways of generating radicals and their functional groups. In the latter review, the authors detailed the synthesis of various peroxidic initiators: di- and oligoperoxy esters, diacyl peroxides, peroxyester-diacyl peroxides or dialkyl peroxides and polyperoxides. Furthermore, polyazoderivatives, azo peroxides and alkyl carbonylazo compounds have been investigated [16,17]. Among them, azo peroxy esters have been the most used [18–20] and their kinetics of decomposition and polymerization reactions have been investigated in detail.

The principle of these syntheses is illustrated in Scheme 1 for the case of a multifunctional initiator [4,16]. From this example, it is clear that the nature of the termination step is crucial for obtaining di-, tri- or multiblock structures. Depending on the type of monomers and on the experimental conditions, recombination or disproportionation reactions are favored.

The efficiency of the process in terms of well-defined structure depends on the efficiency and kinetics of decomposition of the different initiating groups

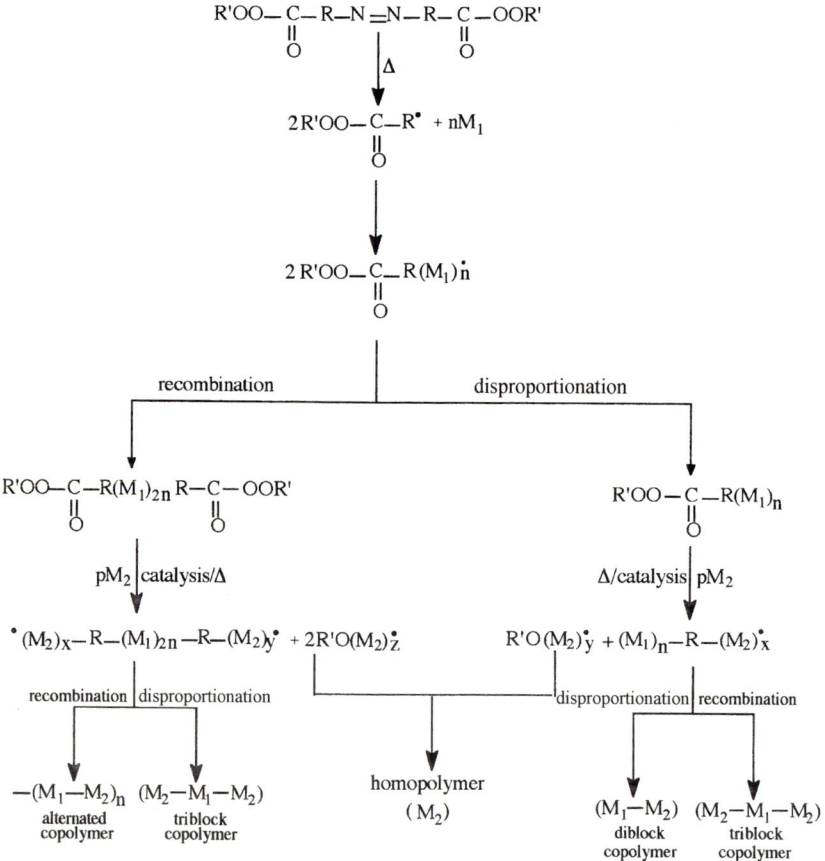

Scheme 1. Decomposition of a multifunctional free radical initiator

included in the initiator. Such studies have been performed in some cases [13, 15, 21] and Table 1 lists several kinetic data relative to the dissociation constants k_{id} of different polyfunctional initiators.

Depending on the types of monomers used and on the predominant mechanism of termination, it is either the azo or the perester group which is decomposed first. This is illustrated by the following examples.

Piirma and Chou [22] synthesized triblock PMMA-b-PS-b-PMMA copolymers according to the following reaction:

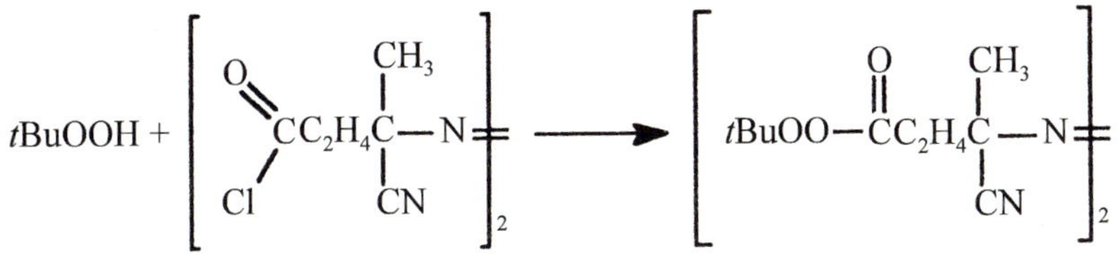

$$tBuOOH + \left[\begin{array}{c} O \\ \diagdown\!\!\!\diagup \\ CC_2H_4C-N\!\!=\!\! \\ \diagup \\ Cl \end{array} \begin{array}{c} CH_3 \\ | \\ | \\ CN \end{array}\right]_2 \longrightarrow \left[\begin{array}{c} O \\ \| \\ tBuOO-CC_2H_4C-N\!\!=\!\! \\ \\ \end{array} \begin{array}{c} CH_3 \\ | \\ | \\ CN \end{array}\right]_2$$

1

$$\mathbf{1} + \xrightarrow{C_6H_5CH=CH_2/60\,°C} \left[\begin{array}{c} O \quad CH_3 \\ \| \quad | \\ tBuOOCC_2H_4C-PS \\ | \\ CN \end{array}\right]_2$$

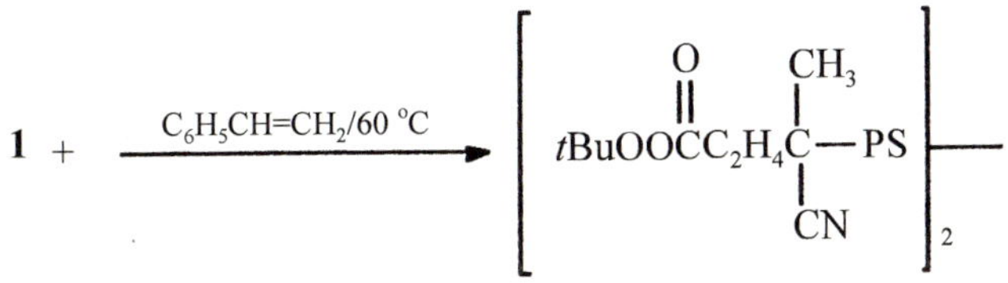

2

$$\xrightarrow[MMA]{25\text{–}30\,°C} PMMA\text{—}b\text{—}PS\text{—}b\text{—}PMMA \,.$$

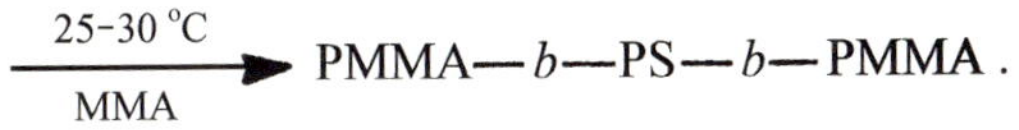

In a first step, the azo function is decomposed at 60 °C, followed by incorporation of PMMA sequences by decomposition of the peroxidic function at 30 °C. The reverse sequence, that is to say, decomposition of perester function followed by that of the azo one, is also possible. For instance, Sheppard et al. patented the synthesis of 42 original azo peresters and polymeric initiators [18,19], some of them being successfully used for the preparation of block copolymers as evidenced by the following examples:

$$\begin{array}{c} H_3C \diagdown \quad N=N \quad \diagup CH_3 \\ \diagup\!\!\!\!\diagdown \quad \diagup\!\!\!\!\diagdown \\ NC \quad C_2H_4 \quad C_2H_4 \quad CN \\ \diagdown \quad \diagup \\ C\text{-}O\text{-}O\text{-}C \\ \| \quad \| \\ O \quad O \end{array} \xrightarrow{Sty\ 50\,°C/24\,h} \left[\begin{array}{c} CH_3 \\ | \\ PS-C_2H_4C-N\!\!=\!\! \\ | \\ CN \end{array}\right]_2$$

$$\xrightarrow[75\,°C/7\,h]{MMA} PS\text{—}b\text{—}PMMA\text{—}b\text{—}PS$$

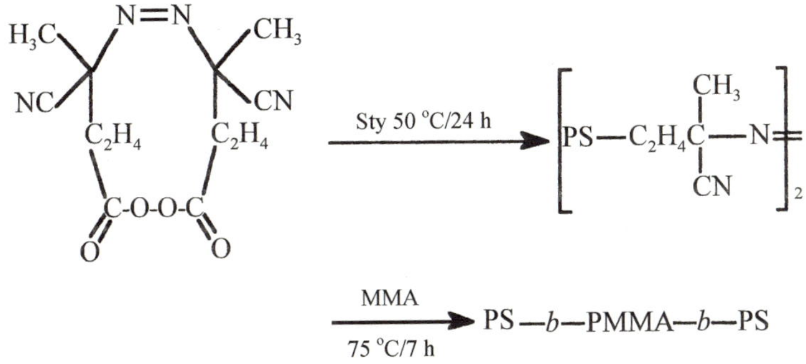

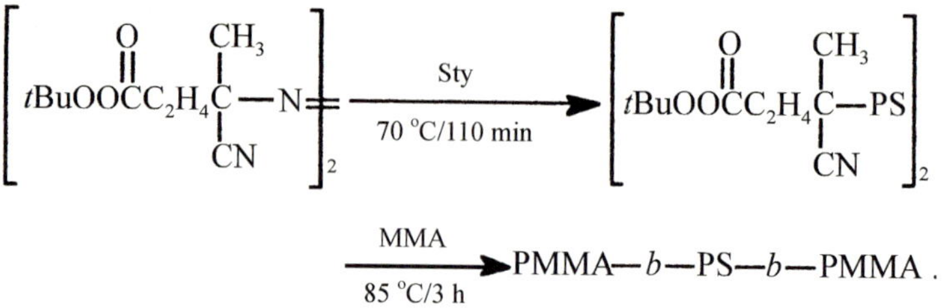

$$[tBuOOCC_2H_4\underset{\underset{CN}{|}}{\overset{\overset{O}{\|}}{C}}-\underset{\underset{}{}}{\overset{\overset{CH_3}{|}}{}}N=]_2 \xrightarrow[70\,°C/110\,min]{Sty} [tBuOOCC_2H_4\underset{\underset{CN}{|}}{\overset{\overset{O}{\|}}{C}}-\underset{\underset{}{}}{\overset{\overset{CH_3}{|}}{}}PS]_2$$

$$\xrightarrow[85\,°C/3\,h]{MMA} PMMA-b-PS-b-PMMA\ .$$

The difference in the decomposition rate constants of the various peroxides or other key groups present in the initiators enables the author to conduct the polymerizations under varying temperature conditions in order to orientate the copolymer structure.

Other procedures can be adequately chosen in order to allow successive decomposition of the initiator. For instance, it is possible to maintain a constant temperature while simultaneously varying reaction times in both decomposition steps, or to activate selectively the thermal decomposition of one group in the presence of specific reactants such as amines, reducing agents or salts of transition metals [16]. Also, selective initiation may be induced by the use of radiation. This assumes the presence on the same initiator of both a thermolabile

Table 1. Kinetic characteristics of polyfunctional initiators (k_{id} is the dissociation constant)

INITIATORS	T°C	$K_{id}\,10^5\,s^{-1}$	T°C	$K_{id}\,10^5\,s^{-1}$
$((CH_3)_3C-O-O-\overset{\overset{O}{\|}}{C}-(CH_2)_2-\underset{\underset{CH_3}{\|}}{\overset{\overset{CN}{\|}}{C}}-N=)_2$	80	12.5	80	3.6
	90	3.8	90	3.8
$((CH_3)_3C-O-O-\overset{\overset{O}{\|}}{C}-O-CH_2-\underset{\underset{CH_3}{\|}}{\overset{\overset{CH_3}{\|}}{C}}-N=)_2$	162	1.9	99	1.9
$(HOO-\underset{\underset{CH_3}{\|}}{\overset{\overset{CH_3}{\|}}{C}}-\underset{\underset{CN}{\|}}{\overset{\overset{CH_3}{\|}}{C}}-N=)_2$	65	1.9	170	1.9
$((CH_3)_3C-O-O-\underset{\underset{CH_3}{\|}}{\overset{\overset{CH_3}{\|}}{C}}-R-\underset{\underset{CH_3}{\|}}{\overset{\overset{CN}{\|}}{C}}-N=)_2$	65	1.9	126	1.9
$((CH_3)_3C-N=N-\underset{\underset{CH_3}{\|}}{\overset{\overset{CH_3}{\|}}{C}}-(CH_2)_3-\overset{\overset{O}{\|}}{C}-O)_2$	76	1.9	63	1.9
$((CH_3)_3C-N=N-\underset{\underset{CN}{\|}}{\overset{\overset{CH_3}{\|}}{C}}-(CH_2)_2-O-\overset{\overset{}{}}{\underset{\underset{O}{\|}}{C}}-O)_2$	78	1.9	45	1.9

group and a photosensitive function or use of two radiation beams (UV source, visible light with the presence of photosensitizers) [16, 17, 23].

Besides the above-mentioned approaches varying the parameters governing the initiator decomposition, it is worth mentioning an interesting biphasic study based on the difference of solubility of monomers and the use an original amphiphilic bifunctional initiator, the palmitoyl (3-carboxy propionyl) sebacoyl diperoxide **3** [24]. In such conditions, block copolymers of water soluble (acrylamide) and oil soluble (butylmethacrylate) monomers can be obtained by interfacial radical copolymerization as follows:

$$yH_2C=CHCNH_2 + xH_2C=C\underset{CO_2Bu}{\overset{CH_3}{\diagup}}$$

(AA)   (BuMA)

$$\downarrow CH_3(CH_2)_{14}-COOCC_8H_{16}COOCC_2H_4CO_2H \quad (3)$$

|——— oil phase ———|—interface—|— H_2O phase ——|

$$CH_3(CH_2)_{14}-\overset{O}{\overset{\|}{C}}-O-(BuMA)_x O\overset{O}{\overset{\|}{C}}C_8H_{16}\overset{O}{\overset{\|}{C}}O-(AA)_y O\overset{O}{\overset{\|}{C}}C_2H_4CO_2H.$$

Both functional groups of the initiator **3** simultaneously generate free radicals to initiate the homopolymerization of both types of monomers in their respective phase. As a result of recombination of macroradicals, amphiphilic block copolymers (in low yield) are formed at the oil-water interface.

2.1.2 Polymeric Initiators

Besides the use of micromolecular multiinitiators, block copolymers can be obtained from macromolecular initiators. In a first step, a polymeric initiator is generally synthesized by reacting a mono- or difunctional polymer with a functional initiator. Various macromolecular initiators were prepared in this way including quite different sequences: polystyrene [13, 18, 19, 25, 26], poly(dimethylsiloxane) [27], poly(methylmethacrylate) [13, 15, 28], polyvinylacetate [28], polyvinylchloride [29, 30], polyesters [30], polycarbonate [31, 32], polybutadiene [13, 25, 33], polyamide [34], poly(ethylene glycol) [35] or polyaromatic [36]. An excellent review of the synthesis and uses of such macroinitiators was written by Nuyken and Voit [37]. Thus, only few typical examples are going to be mentioned below.

In their patents, Sheppared et al. [18,19] proposed the synthesis of a series of di- or triblock copolymers from polyazomacroinitiators including various types of sequences: polystyrene, polybutadiene, polyamide, poly(butyl sebacate), polyether, poly(butyl azelate) or polycarbonate. They observed that these macroinitiators usually decomposed at higher temperatures than that of AIBN except for ω-t-butyl azo macroinitiators. A few years later, Campbell et al. [25, 26] performed the syntheses of a hard-b-soft copolymer which constitutes a structure of great technological interest. They prepared a multiblock polybutadiene-b-polystyrene (PS-b-PBd) copolymer by condensation of hydroxytelechelic PS with an azo initiator followed by sequencial addition of butadiene:

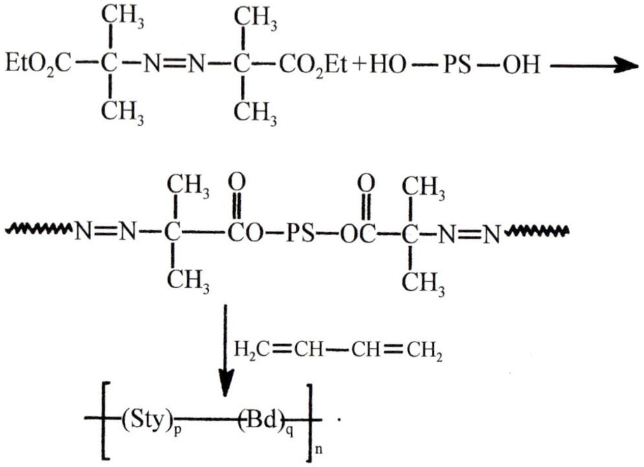

$$EtO_2C-\underset{\underset{CH_3}{|}}{\overset{\overset{CH_3}{|}}{C}}-N=N-\underset{\underset{CH_3}{|}}{\overset{\overset{CH_3}{|}}{C}}-CO_2Et + HO-PS-OH \longrightarrow$$

$$\sim\sim\sim N=N-\underset{\underset{CH_3}{|}}{\overset{\overset{CH_3}{|}}{C}}-\overset{\overset{O}{\|}}{C}O-PS-O\overset{\overset{O}{\|}}{C}-\underset{\underset{CH_3}{|}}{\overset{\overset{CH_3}{|}}{C}}-N=N\sim\sim\sim$$

$$\downarrow H_2C=CH-CH=CH_2$$

$$\left[-(Sty)_p - (Bd)_q - \right]_n$$

Using a different approach, Hazer and Kurt [38] achieved the synthesis of polystyrene grafted copolymers with polybutadiene backbones. The method includes the synthesis of a macroperoxide initiator containing polystyrene which further reacts on the double bonds of PBd.

Amphiphilic triblock copolymers were prepared from poly(ethylene glycol) (PEG) [35]. Starting from PEG segments of different lengths (Mw = 200, 400, 600, 1000 and 1500) and with different end-groups (e.g. hydroxy, acetoxy, benzoyl, oleoyl, phenylurethane), telechelic macroazoinitiators are first prepared. By polymerization of styrene, a wide range of original copolymers are produced:

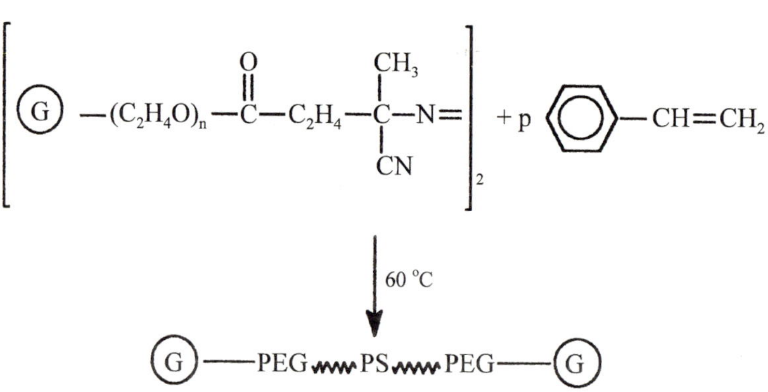

$$\left[\textcircled{G} -(C_2H_4O)_n-\overset{\overset{O}{\|}}{C}-C_2H_4-\underset{\underset{CN}{|}}{\overset{\overset{CH_3}{|}}{C}}-N= \right]_2 + p \; \text{C}_6\text{H}_5-CH=CH_2$$

$$\downarrow 60\ °C$$

$$\textcircled{G}-PEG\sim\sim\sim PS\sim\sim\sim PEG-\textcircled{G}$$

with $\textcircled{G}$: HO, $C_6H_5CO_2$, CH_3CO, $C_6H_5NHCO_2 - NHCO_2$, $C_{18}H_{37}O$

and n = 4, 8, 13, 22 and 33.

Whatever the functional end-group, these block copolymers are obtained in high yields and are soluble in most common organic solvents. They differ in their molecular weights and their thermal properties.

Polymeric radical initiators can also be generated by reacting a living ionic polymer with α, ω-difunctionalized organic radical initiators [39]. These living polymers are prepared by an anionic or cationic process. Examples of the use of anionically prepared precursor are numerous. Starting from a macro-initiator produced by anionic living polystyryl or polyisoprenyl polymers with either a peroxy alkyl halide [40], an azo group [41, 42] or molecular oxygen [43, 44], block copolymers are prepared according to the following scheme:

$$\sim\sim\sim (M_1)_{n-1}M_1^- Met^+ + R-Q-Q-R \longrightarrow$$

$$\sim\sim\sim (M_1)_n-R-Q-Q-R(M_1)_n \sim\sim\sim$$
$$\mathbf{4}$$

$$\mathbf{4} \xrightarrow{\Delta} \sim\sim\sim (M_1)_n \sim\sim\sim R-Q^\cdot$$
$$\mathbf{5}$$

$$\mathbf{5} \xrightarrow{M_2} \sim\sim\sim (M_1)_n-R-Q(M_2)_p QR(M_1)_n \sim\sim\sim$$

$$+ \sim\sim\sim (M_1)_n-RQ(M_2)_p \cdot$$

In this scheme, Met represents a metal and R-Q-Q-R the initiator (R is a halogen atom or a CN group and Q a diazo or peroxide group). The authors discuss the influence of the instability of C-Met bonds formed with different metals (silver, lead, mercury, iron or copper) [39, 45]. Interestingly, in the case of the C–Hg bond, the reaction is as follows:

$$\sim\sim\sim M_1^- Li^+ + HgX_2 \longrightarrow \sim\sim\sim M_1-HgX + LiX$$

$$\sim\sim\sim M_1-HgX \xrightarrow{\Delta} \sim\sim\sim M_1^\cdot + 1/2\, Hg_2X_2$$

$$\sim\sim\sim M_1^\cdot + pM_2 \longrightarrow \sim\sim\sim (M_1)_n-(M_2)_p \sim\sim\sim \cdot$$

In this case, the formation of ($\sim\sim\sim M_1-Hg-M_2 \sim\sim\sim$) bonds by Wurtz coupling can be minimized by using an excess of mercury salt.

Precursors prepared by the cationic living method and mostly poly(THF) precursors were also proposed. In this way, from poly(THF) containing peroxide or azo groups prepared by reaction with either alkali salt of succinic acid peroxide [45], LiOOCH$_2$Br [46] or α,ω-diacid chloride [47], they lead to interesting block copolymers.

It is seen that free radical micromolecular or macromolecular initiators have been successfully employed for the synthesis of di-, tri- or multiblock copolymers. However, once again, the structure of these block copolymers depends upon the termination step of the polymerization, and especially on the recombination or disproportionation of macroradicals produced. Besides, such a method also generates homopolymers. Separation and purification of these different structures are usually very difficult or even impossible. Moreover, the copolymers obtained usually exhibit a broad polydispersity, a defect inherent in the classical radical process.

To date, many new techniques have been proposed and developed to control the reactivity of the radicals. Such control may give a "living" character to the radical polymerization. While the first example of living polymerization was introduced in 1956 [48] in anionic polymerization, it is only in 1982 with the pioneer work of Otsu et al. [49, 50] that the possibility of having a "living" process in free radical polymerization was demonstrated. Otsu, using thiuram disulfide compounds, developed the Iniferter concept with control of chain termination (see Sect. 3.2.4). Later on, Solomon et al. [51] and Georges et al. [52] proposed the use of stable counter radicals as thermally labile capping agents for the growing polymeric chain. This very promising approach is nowadays the subject of active research and allows in principle the preparation of block copolymers with narrow molecular weight distributions in one pot, as with ionic methods, but without high purity and vacuum requirements. However, despite considerable progress, the truly living character is still to be attained and it seems preferable to use the term controlled process rather than living process.

2.2 Controlled Radical Polymerization Involving Stable Free Radicals

The counter radical process is based on the reversible capture of the carbon centered radicals formed by reaction of an initiating radical with monomer before propagation can occur. Upon heating or UV exposure, the resulting product can be homolytically cleaved in such a way that another monomer unit, or a number of monomer units, add before the macroradical is again captured. The process is repeated until no monomer remains or irreversible termination occurs.

Two main types of counter radicals are proposed, purely organic (•O-X) issued from nitroxyls, alkoyamines and arylazooxyls, and organometallic complexes (**C**) able to generate stable free radicals. Equilibria may be written as follows:

$$I-M_n^{\bullet} + {}^{\bullet}O-X \underset{UV}{\overset{\Delta}{\rightleftharpoons}} I-M_n-O-X$$

$$I-M_n-M^{\bullet} + C \underset{UV}{\overset{\Delta}{\rightleftharpoons}} I-M_n-(MC)^{\bullet}.$$

Difficulties arise from the various possible interactions between the counter radical, the initiator and the monomer. The truly living character is only demonstrated when some requirements are fulfilled. Molecular weight must increase in a linear fashion with conversion. Polydispersity must be narrow and lower than that in the classical process (theoretical value, Mw/Mn = 1.5) which assumes, in particular, a rapid initiation step. The obtaining of high and strictly controlled molecular weight must be possible. Concerning the preparation of block copolymers, this also assumes that each sequence is of controlled and uniform length, a requirement which is not always easy to demonstrate.

The counter radical method has been studied with various monomers more or less successfully. However, the synthesis of only few block or grafted copolymers is effectively described. This is a strong indication that a true control of the polymerization is still not achieved with all monomers although progress is constant. Nevertheless, it is clear that the possibility of reversibly controlling the termination step offers a tool of choice for the synthesis of well-defined and pure block copolymers and many studies are still necessary to understand properly the precise mechanism of macroradical end capping in order to control the reversible character and possible secondary reactions.

2.2.1 Nitroxyls and Related Compounds

Alkoxyamines have been used successfully in controlled radical polymerization and they are most used as persistent radicals. They can be produced by at least three methods: by oxidation of the corresponding hindered amines [53, 54], from the reaction between free radicals and nitrone, or from the photolysis of amines [55]. They can be either added to the polymerization medium or produced in situ [51, 52, 56–58]. They are stable at room temperature and can be isolated. Their general structure is $C-O-NR_1R_2$ in which R_1 and R_2 represent alkyl or aryl groups usually exhibiting important steric constraints, but not a hydrogen atom. They can be used as initiators with the great advantage that the nitroxide formed by cleavage of the C–O bond appears inert towards many monomers and does not initiate radical polymerization. The lability and rate of homolysis of the C–ON bond strongly depends on the substituents on the carbon and nitrogen atoms, i.e., on the monomer and nitroxyls types, and on the experimental conditions [59, 60]. Different nitroxides have been studied and several methods have been proposed to increase the lability of the C–ON bond in order to perform the polymerizaton at sufficiently low temperatures and to shift the position of the equilibrium between free and capped macroradicals [61, 62]. The most used stable nitroxyl is the 2,2,6,6-tetramethyl piperidinyl-1-oxyl (Tempo). However, in the polymerization of styrene, temperatures above 100 °C are required in order to obtain a sufficient rate of monomer insertion. As a disadvantage, self initiation of styrene proceeds via Diels-Alder reaction [62]. It must also be emphasized that the effectiveness of the method depends strongly on the structure of the monomer and that H abstraction or β-elimination are often observed [63–67]. In particular, with acrylic monomers only low molecular weights are produced [68, 69].

In their patent, Solomon et al. [51], using alkoxyamine, described the first controlled synthesis of PMA-*b*-PS and PMA-*b*-PEA (Mn = 4300 and Mw/Mn = 1.7) block copolymers from methyl acrylate macroradicals (Mn = 2500 and Mw/Mn = 1.5). The reactions are as follows:

$$\underset{R_2}{\overset{R_1}{\diagdown}}NO-R_3 + H_2C=CH(CO_2CH_3) \longrightarrow PMA-ON\underset{R_2}{\overset{R_1}{\diagup}}$$

$$\swarrow Sty \qquad \searrow EA$$

$$PMA-b-PS \qquad PMA-b-PEA$$

The same group proposes the preparation of PEA-*b*-PMA copolymers able to react further over methyl methacrylate to yield PEA-*b*-PMA-*b*-PMMA triblock copolymers (Mw = 10500 and Mw/Mn = 2.6) [68, 69]. Using indoline oxyl radical, PEA with narrow polydispersity was first prepared and, under a certain temperature (120 °C), the ω-indolinoxyl-PEA produces a PEA radical able to polymerize MA, and, in a third step, MMA

An other interesting example of copolymer is given by Georges et al. [52, 59] who first demonstrated the living character of the polymerization of styrene initiated by dibenzoyl peroxide in the presence of Tempo or Proxyl (2,2,5,5-tetramethyl-1-pyrrolydinyloxy). Polystyrene with a narrow polydispersity (Mw/Mn = 1.2) is obtained and block copolymers with butadiene, isoprene, acrylate and methacrylate sequences are prepared:

$$\text{Styrene} + [C_6H_5CO_2]_2 + \underset{R_2}{\overset{R_1}{\diagdown}}N-O^{\bullet} \xrightarrow[\text{or suspension}]{\text{bulk, solution}} PS-CH_2CH(C_6H_5)-ON\underset{R_2}{\overset{R_1}{\diagup}}$$

6   **7**

$$7 \rightleftharpoons [PS-CH_2\overset{\bullet}{C}H(C_6H_5) \quad {^\bullet}ON\underset{R_2}{\overset{R_1}{\diagup}}] \xrightarrow{\text{butadiene}} PS-b-PBd.$$

Polydispersity of the copolymer is still narrow (Mw/Mn = 1.36) whereas it increases to 4.21 in the absence of Tempo, leading to a bimodal molecular weight distribution. Such a process is also applied in aqueous medium [70] for the polymerization of styrenesulfonic acid sodium salt with narrow molecular weight distribution (Mw/Mn as low as 1.18).

Recent investigations in our laboratory involve the controlled radical polymerization of styrene, initiated by dicumyl peroxide **8** in the presence of Tempo [71, 72]. Molecular weights are obtained in the range 2500–300 000 with narrow polydispersity (Mw/Mn = 1.5). From such capped Tempo polystyrene **9**, polystyrene-*b*-chloromethylstyrene (PS-*b*-PCMSty) [171], PS-*b*-PBd and PS-*b*-PBd-*b*-PS block copolymers are prepared with Mw = 50 000 and Mw/Mn = 1.5 [72]:

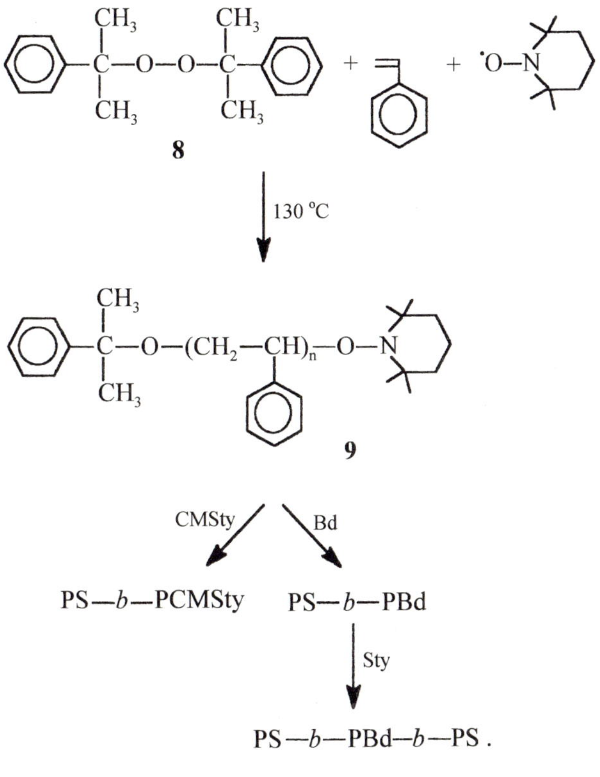

$$\text{Ph-C(CH}_3\text{)}_2\text{-O-O-C(CH}_3\text{)}_2\text{-Ph} \quad + \quad \text{styrene} \quad + \quad \text{·O-N(TEMPO)}$$

8

$$\downarrow 130\,°C$$

$$\text{Ph-C(CH}_3\text{)}_2\text{-O-(CH}_2\text{-CH(Ph))}_n\text{-O-N(TEMPO)}$$

9

$$\swarrow \text{CMSty} \qquad \searrow \text{Bd}$$

PS—*b*—PCMSty   PS—*b*—PBd

$$\downarrow \text{Sty}$$

PS—*b*—PBd—*b*—PS .

Another interesting approach developed by Matyjaszewski et al. [62, 73, 74] is based on the use of nitroxyl radicals in the presence of organoaluminum complexes in association with various ligands in order to activate the homolytic cleavage of the counter radical. Applied to vinyl acetate (VAc) or MMA, the polymers produced undergo a further stepwise polymerization of styrene, MMA or VAc to form PVAc-*b*-PS, PVAc-*b*-PMMA and PMMA-*b*-PVAc copolymers. Average number molecular weights ranged from 5000 to 23 000 with polydispersity indexes varying from 1.16 to 1.48 [73]. The complex system involved was as follows:

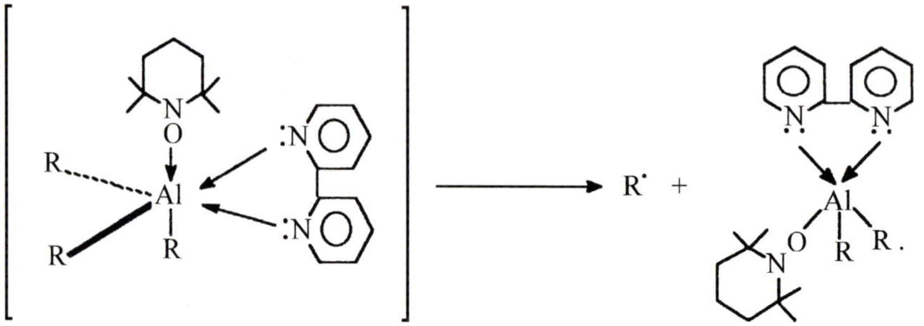

$$\left[\begin{array}{c} \text{(piperidinyl-N-O)Al(R)(R)} \leftarrow \text{:N(pyridyl)} \\ \text{:N(pyridyl)} \end{array} \right] \longrightarrow R^\bullet + \text{(bis-pyridyl-Al(O-piperidinyl)(R)(R))}$$

with R = iBu

Although alkoxyamines were mainly used, a number of other initiators was studied [57, 58, 75–77]. Druliner [57] describes a series of electron-transfer initiators able to generate long-lived oxygen-centered radicals such as **10** which are associated with the growing end of the acrylate and methacrylate chains. Nearly pure block copolymers are prepared from sequential polymerization of MA, BuA and MMA:

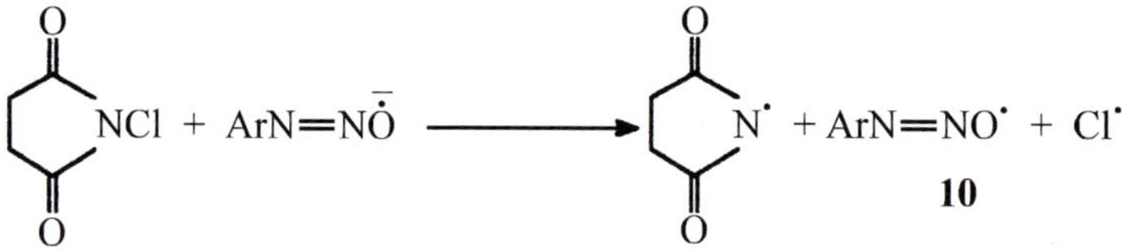

$$\text{(succinimidyl)NCl} + \text{ArN}=\overset{-}{\text{NO}} \longrightarrow \text{(succinimidyl)N}^\bullet + \text{ArN}=\text{NO}^\bullet + \text{Cl}^\bullet$$
                                                                                                                          10

$$\mathbf{10} \xrightarrow[\text{BrC(CH}_3)_2\text{CO}_2\text{Et}]{\text{MMA}} \text{PMMA}-\text{ON}=\text{NAr} \xrightarrow{\text{BuA}} \text{PMMA}-b-\text{PBuA}.$$

The reaction was carried out at 25–60 °C in different solvents and leads to high molecular weights although with polydispersity indexes higher than 3 based on SEC determinations. Recently, Yamada et al. [75] studied initiators derived from the triphenylverdazyl radical. PS-b-PMMA copolymers are prepared although with low yields.

The counter radical method can also be used for graft copolymer synthesis. Solomon et al. propose two routes [51]. The first one involves copolymerization with a functional monomer such as methacrylate containing pendant alkoxyamine. In the second route, the alkoxyamine is grafted onto a polymer precursor used in a second step to initiate the living polymerization of a second monomer. PBd-g-PMA is prepared this way from PBd.

Another method is proposed by Hawker et al. [78]. The principle is illustrated with the synthesis of grafted copolymer PS-g-PS:

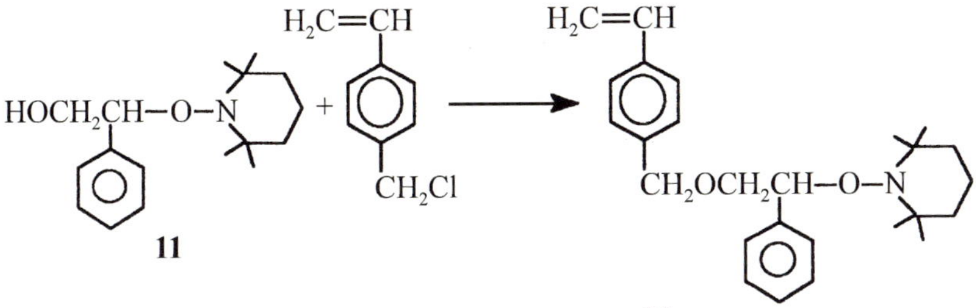

[Scheme showing compounds 11, 12, 13, 14 with reactions]

11 + p-chloromethylstyrene → 12

12 —AIBN/Sty→ 13

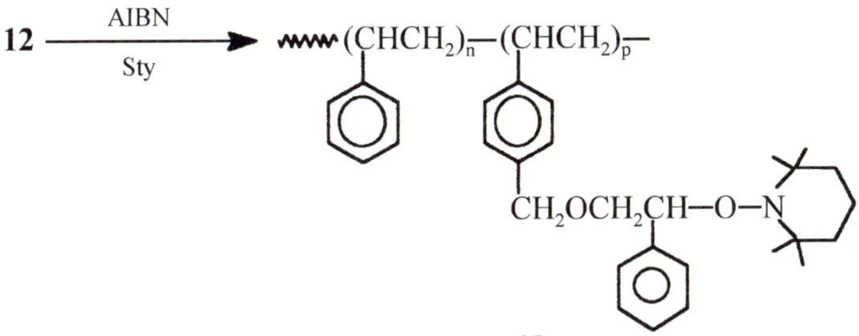

13 —Sty 130 °C→ PS—g—PS (14)

The hydroxy-tempo derivative **11** is first reacted with *p*-chloromethyl styrene to give a Tempo capped polymerizable styrenic compound **12**. Copolymerization of **12** with styrene gives the multifunctional initiator **13**, which has a PS backbone with attached Tempo groups. Reaction of **13** with styrene at 130 °C gives the grafted copolymer **14**. After cleavage of the benzyl ether bonds, a Mn of 23 000 is determined with a Mw/Mn value of 1.20.

2.2.2 Organometallic Complexes

Coordination chemistry has become a powerful tool for the control and the living nature of radical polymerization [79, 80]. Various examples show that the role of initiator and counter radical can be played by organometallic species with an even number of electrons. Besides aluminum complexes used by Matyjaszewski, several other transition metals, metallocenes, and organolanthanides with various ligands have been studied in controlled radical polymerization [79–97]. In some cases, a controlled polymerization was achieved [81, 83–85, 87, 90–94, 97]. However, the mechanism of the polymerization is not always known and it may happen that heterolytic cleavage of the active bond occurs in competition with the homolytic cleavage. In these conditions, a pure radical process cannot be assumed and only few block copolymers have been prepared.

Examples of preparation of copolymers are scarce. Mun et al. [81, 82] showed that the binary system of cobaltocene/bis(ethylacetoacetato) copper (II) effectively initiates the living radical polymerizaton of MMA at 25 °C in acetonitrile. The polymerization activity of this initiator system was markedly affected by the solvent used. The synthesis of PMMA-*b*-PS copolymers with molecular weights reaching 700 000 was successfully attempted by adding styrene to the living PMMA. The yield of the copolymers reached 80% when the MMA polymerization was carried out for three days. The same team [91] also synthesized PS-*b*-PMMA copolymers from the polymerization of MMA with polystyrene obtained in the presence of reduced nickel/halide systems. The yields range from 84 to 91% depending on the halide complex used.

In addition, Arvanitopoulos et al. [84] achieved the synthesis of di- and triblock copolymers of acrylates using cobalt oxime complexes and telechelic initiators (e.g., 1,6-[di(pyridinato)cobaltoxime]hexane), respectively.

Organoborane complexes have shown a high efficiency for polymerizing vinyl monomers at room temperature as evidenced by numerous investigations developed by Chung et al. [94–97]. The oxidation of organoborane leads to the formation of borane peroxide able to initiate the polymerization:

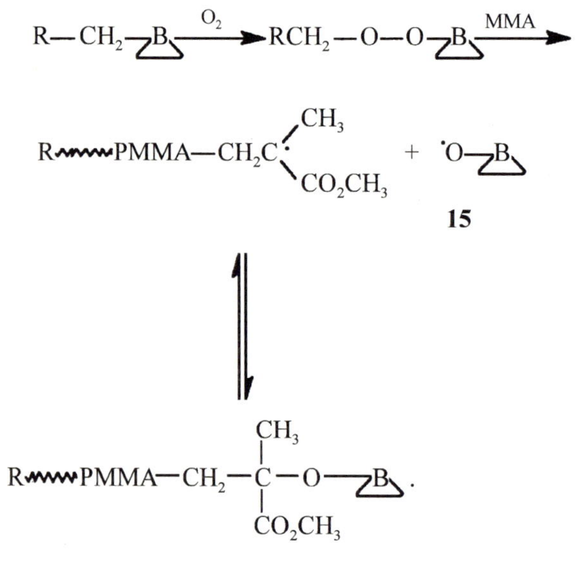

$$R-CH_2-B \xrightarrow{O_2} RCH_2-O-O-B \xrightarrow{MMA}$$

$$R\text{\textasciitilde}PMMA-CH_2C\begin{smallmatrix}CH_3\\CO_2CH_3\end{smallmatrix} + \cdot O-B$$

$$\mathbf{15}$$

$$\updownarrow$$

$$R\text{\textasciitilde}PMMA-CH_2-\underset{CO_2CH_3}{\overset{CH_3}{C}}-O-B \, .$$

The oxyborane radical **15** acts as stable counter radical and assures the reversible termination. Molecular weights up to 150 000 with a polydispersity of 2.4 were obtained. Such a system was also satisfactorily used by Jiang et al. [98]

with telechelic diborane for the synthesis of PS-*b*-PMMA, PS-*b*-PHEMA, PS-*b*-poly(ε-caprolactone) and PMMA-*b*-PCL block copolymers:

$$H_2C=CHCH_2CH_2CH=CH_2 \xrightarrow[RT]{HBR_1R_2} H_2C=CH(CH_2)_4B\begin{smallmatrix}R_1\\R_2\end{smallmatrix}$$

$$\xrightarrow[Sty/THF]{O_2} H_2C=CH(CH_2)_4-(Sty)_nB\begin{smallmatrix}R_1\\R_2\end{smallmatrix}$$
16

$$\mathbf{16} \xrightarrow[RT]{HBR_1R_2} R_1R_2B(CH_2)_6(Sty)_nBR_1R_2 \xrightarrow[MMA/THF]{O_2}$$

$$R_1R_2B(MMA)_p(CH_2)_6(Sty)_nBR_1R_2 .$$

2.3 Conclusion

New approaches based on the introduction of reactive species into reaction mixtures that tend to cap the growing chains reversibly allow, in many cases, production of well-defined polymers and copolymers with narrow polydispersities. Up to few years ago, such a possibility was unobtainable by a classical free radical process. The proposed principle of control of macroradical reactivity is very interesting conceptually, and represents a very powerful tool to prepare block copolymers with well-controlled structures. However, it is clear that the true living character as demonstrated by some anionic polymerizations is still not obtained and much more work needs to be done to understand and control this new process better.

3 Synthesis of Block Copolymers by Radical Telomerization

Radical telomerization is an efficient method for preparing block cotelomers, block cooligomers or block copolymers, as evidenced by numerous investigations in this area. In this section, traditional telomerization which utilizes a transfer agent, a monomer and a generator of free radicals, is first going to be discussed for the synthesis of non-living block copolymers. More recent results concerning the obtaining of other block copolymers by pseudo-living or dormant telomerization will be mentioned later.

3.1 Uncontrolled Radical Telomerization

Two separate topics must be considered: the block copolymers produced from bistelomerization of two different monomers with a telechelic or a difunctionalizable telogen and the coupling of monofunctional telomers.

3.1.1 Bistelomerization

The basic principle consists of introducing a specific group at one extremity of a polymeric chain made of a first monomer and this macromolecular species then being able to initiate the polymerization of a second monomer. Such an end-group is generated either directly in the first polymerization or by subsequent chemical change of one or both chain-extremities.

3.1.1.1 Direct Method

Bamford is considered as the pioneer of this research. His investigations led to a series of various publications dealing with the synthesis of block and graft copolymers [99]. The following example [100] using the ω-tribromomethyl polymer produced from the radical telomerization of a conventional monomer M_1 with the tetrabromomethane illustrates this kind of polymerization:

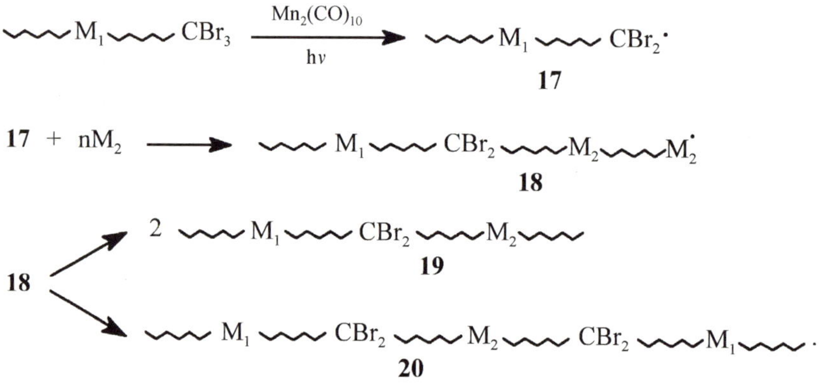

$$\sim\!\!\sim\!\! M_1 \!\sim\!\!\sim\! CBr_3 \xrightarrow[h\nu]{Mn_2(CO)_{10}} \sim\!\!\sim\!\! M_1 \!\sim\!\!\sim\! CBr_2^{\bullet}$$
$$\mathbf{17}$$

$$\mathbf{17} + nM_2 \longrightarrow \sim\!\!\sim\! M_1 \!\sim\!\!\sim\! CBr_2 \!\sim\!\!\sim\! M_2 \!\sim\!\!\sim\! M_2^{\bullet}$$
$$\mathbf{18}$$

$$\mathbf{18} \begin{cases} 2 \sim\!\!\sim\! M_1 \!\sim\!\!\sim\! CBr_2 \!\sim\!\!\sim\! M_2 \!\sim\!\!\sim \\ \qquad\qquad\qquad \mathbf{19} \\ \sim\!\!\sim\! M_1 \!\sim\!\!\sim\! CBr_2 \!\sim\!\!\sim\! M_2 \!\sim\!\!\sim\! CBr_2 \!\sim\!\!\sim\! M_1 \!\sim\!\!\sim \\ \qquad\qquad\qquad \mathbf{20} \end{cases}$$

M_1 and M_2 represent either methyl acrylate (MA), styrene (Sty), butadiene (Bd) or isoprene (Ip), or are already polymeric chains composed of alternating Sty-*alt*-MA, Ip-*alt*-MMA, Sty-*alt*-MMA, Bd-*alt*-MMA, Ip-*alt*-MA, Bd-*alt*-MA [99].

In this specific work, the authors have shown that the chain-termination mainly occurred by recombination. This explains the obtaining of A-*b*-B-*b*-A triblocks copolymers. Similar investigations were performed by Niwa et al. [101] who prepared amphiphilic PEA-*b*-PS copolymers with controlled block length by stepwise polymerization of ethyl acrylate and styrene from tetrabromomethane. In the same way, Kang et al. [102] synthesized PS-*b*-PMMA copolymers.

However, the evidence of the above mechanism proposed by Bamford is not fully convincing since the authors have not taken into account the process of transfer inherent to the redox systems. Such a process was enhanced by Asscher and Vofsi [103] for $FeCl_3$/benzoin system and also by Freidlina and Chukovskaya [104] for carbonyl metals close to those used by Bamford:

$$Fe^{3+} + benzoin \longrightarrow Fe^{2+} + benzyle$$

$$Fe^{2+} + CCl_4 \longrightarrow FeCl_3 + CCl_3^{\cdot}$$

$$CCl_3^{\cdot} + nM \longrightarrow CCl_3-(M)_n^{\cdot}$$

$$Cl_3C-(M)_n^{\cdot} + FeCl_3 \longrightarrow CCl_3-(M)_n-Cl + FeCl_2.$$

In our group, this above mechanism was kinetically demonstrated [105,106] just like the general equation in which the average degree of polymerization (or telomerization), DP_n, depends on the characteristic ratios $C_{Me} = [Fe^{2+}]_0/[M]_0$ and $R_0 = [CCl_4]_0/[M]_0$, where $[M]_0$ represents the initial monomer concentration (ϕ designates a corrective parameter linked to the higher catalytic behaviour of Fe^{2+} rather than that of Fe^{3+}, and C_{CCl_4} represents the transfer constant of CCl_4):

$$\frac{1}{\overline{DP_n}} = C_{Me} \frac{\phi[Fe^{3+}]}{[M]} + C_{CCl_4} \frac{[CCl_4]}{[M]}.$$

The first term is usually very high and it demonstrates perfectly the step of termination/redox transfer. Under these conditions, it is believed that Bamford observed such a process in his investigation since the value C_{Me} is 75 and C_{CCl_4} 0.013. From such a mechanism, numerous bistelomerizations were performed in our laboratory to produce either monoadducts or diblock cotelomers [107,108]. Thus, difunctionalization by allyl acetate of telomers which exhibit trichloromethyl end-groups was successfully achieved as follows:

$$CCl_3-(CF_2-CFCl)_n-CF_2-CCl_3 + CH_2=CH-CH_2-OAc \longrightarrow$$

$$AcO-CH_2-CHCl-CH_2-CCl_2-(CF_2-CFCl)_n-$$

$$CF_2-CCl_2-CH_2-CHCl-CH_2-OAc.$$

This is similar to the synthesis of macromonomers of vinyl chloride [109]. Furthermore, the following photocrosslinkable diblock cooligomers were synthesized [110] from the similar system used by Bamford but using CCl_4 as the linking group of both blocks:

$$Cl-(M)_n-CCl_2-(CH_2-CH)_p-Cl$$
$$\qquad\qquad\qquad\qquad\quad |$$
$$\qquad\qquad\qquad\quad CO_2-CH_2-CH-CH_2-O_2C-CH=CH_2$$
$$\qquad\qquad\qquad\qquad\qquad\qquad\quad |$$
$$\qquad\qquad\qquad\qquad\qquad\qquad\quad OH$$

with $M = CFCl-CF_2$ or $CH_2-CH=C-CH_2$.
$$\qquad\qquad\qquad\qquad\qquad\qquad\;\; |$$
$$\qquad\qquad\qquad\qquad\qquad\qquad\; CH_3$$

An example of our synthesis of diblock cooligomers bearing about 100 units of two different monomers by a stepwise process involving redox and then radical initiations was described in 1978 [111]:

$$CCl_4 + nCH_2=CH \xrightarrow[CH_3CN]{FeCl_3/benzoin} CCl_3-(CH_2-CH)_n-Cl$$
$$\qquad\qquad\quad |\qquad\qquad\qquad\qquad\qquad\qquad\quad |$$
$$\qquad\qquad\; CO_2R\qquad\qquad\qquad\qquad\qquad\quad CO_2R$$
$$\qquad\qquad\qquad\qquad\qquad\qquad\mathbf{21}$$

$$\mathbf{21} + pCH_2=CH \xrightarrow[peroxide]{benzoyl} Cl-(CH-CH_2)_n-CCl_2-(CH_2-CH)_p-Cl.$$
$$\qquad\qquad\quad |\qquad\qquad\qquad\qquad |\qquad\qquad\qquad\qquad |$$
$$\qquad\qquad\; OAc\qquad\qquad\qquad\quad CO_2R\qquad\qquad\qquad\; OAc$$

A similar example using two radical initiations was published in 1980 by a Korean team [112], leading to a PS-*b*-PMMA cooligomer:

$$Cl-(Sty)_n-CCl_2-(MMA)_p-Cl.$$

Such a direct method of bistelomerization to prepare diblock cooligomers is very interesting since the kinetics allows one to predict the length of both blocks in an accurate way. However, it requires a very good knowledge of both the systems of redox and radical initiations and also of the ability of transfer of species introduced in chain-ends.

3.1.1.2 Indirect Method

The indirect method involves the preparation of a first block followed by a chemical change to introduce a reactive end-group. This reactive group allows the initation of the second growing-chain. In this case, the first block can be obtained by various routes of macromolecular synthesis: polycondensation and ionic or radical polymerization.

Thus, by polycondensation of H_2S with α,ω-dimethacrylate-PEG, the following telechelic macrothiols have been produced [113,114]:

$$HS-\left[CH_2-C(CH_3)H-CO_2-(C_2H_4O)_n-CO-CH(CH_3)-CH_2-S\right]_p-H$$

$n = 3 - 100 \qquad p \leq 100$.

These α,ω-dithiol polymers have been used for the initiation of methacrylates with fluorinated side groups such as

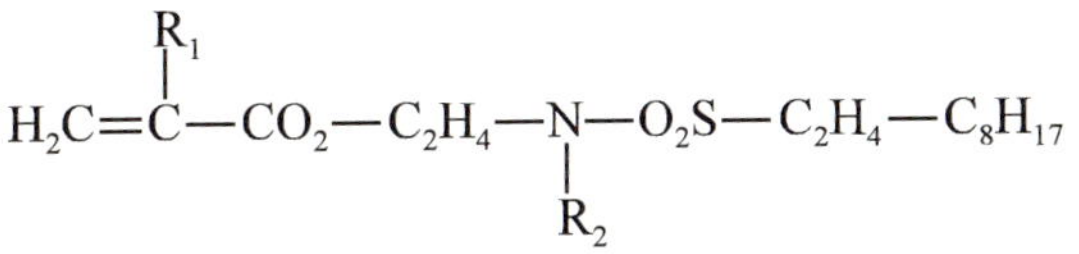

$$H_2C=\underset{R_2}{\overset{R_1}{\underset{|}{\overset{|}{C}}}}-CO_2-C_2H_4-N-O_2S-C_2H_4-C_8H_{17}$$

with R_1 = H or CH_3 and $R_2 = CH_3$ or C_2H_5.

A novel triblock cooligomer in which both hydrophobic blocks are separated by a hydrophilic sequence was thus obtained. Such a cooligomer commercialized by 3M under the Scotchguard trademark is still successfully used for textile treatment [115]. Similar investigations were also performed on the telomerization of the corresponding fluoroacrylates with mercapto terminated polyethylene glycols prepared by condensation of PEG with thioglycolic acid [116]. In this case, a diblock copolymer useful for textile coatings and called Foraperle was produced by Elf Atochem as follows:

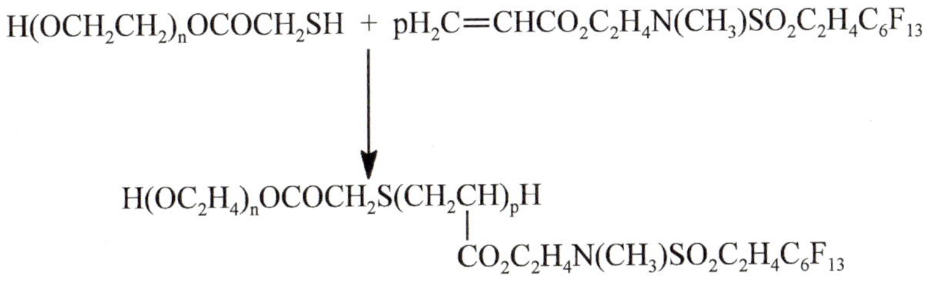

$$H(OCH_2CH_2)_nOCOCH_2SH + pH_2C=CHCO_2C_2H_4N(CH_3)SO_2C_2H_4C_6F_{13}$$

$$\downarrow$$

$$H(OC_2H_4)_nOCOCH_2S(CH_2CH)_pH$$
$$|$$
$$CO_2C_2H_4N(CH_3)SO_2C_2H_4C_6F_{13}$$

where $n = 13-14$.

A similar strategy was reported by Asahi Glass which prepared fluoroacrylic (FA) diblock copolymers used as surface active agents [117]. Acrylates containing perfluorinated groups were telomerized with mercapto acetic acid followed by the condensation of the telomer produced with polypropylene oxide (PPO) to yield PPO-b-PFA diblock copolymers.

As an example of first block obtained by ionic polymerization, Tung et al. [118] have synthesized an α,ω-polybutadiene, the end-groups of which were deactivated onto episulfide in order to generate a telechelic polybutadiene dithiol. Such a compound was successfully used to initiate the copolymerization of styrene and acrylonitrile to form thermoplastic elastomers containing a central polybutadiene sequence and lateral styrene-acrylonitrile blocks. In the same way, the deactivation of an anionic polystyrene with dibromomethyl benzene leads to an ω-$BrCH_2C_6H_4CH_2$-polystyrene which allows the initiation of MMA according to a redox process involving $Mn_2(CO)_{10}$ [119] as described by Bamford [99] and Bamford and Han [100].

An example of radical polymerization with chemical change was suggested by Sato et al. [120] who used thioacetic acid as the transfer agent:

$$CH_3COSH + nCH_2{=}CH(OAc) \xrightarrow{AIBN} CH_3{-}CO{-}S{-}(CH_2{-}CH(OAc))_n{-}H$$
$$\mathbf{22}$$

$$\mathbf{22} \xrightarrow{NaOH/EtOH} HS{-}(CH_2{-}CH(OH))_n{-}H$$
$$\mathbf{23}$$

$$\mathbf{23} \xrightarrow[H_2O/KBrO_3]{Acrylic\ acid} H{-}(CH(CO_2H){-}CH_2)_p{-}S{-}(CH_2{-}CH(OH))_n{-}H \ .$$

An extension to telechelics was described almost 30 years ago by Uraneck [121] who reacted xanthogen disulfide with butadiene, leading to an original styrene-butadiene-styrene triblock copolymer formed in a radical way as follows:

$$ROCS(=S){-}S(S=)COR + nH_2C{=}CH{-}CH{=}CH_2 \longrightarrow ROCS(=S)(Bd)_nS(S=)COR$$
$$\mathbf{24}$$

$$\mathbf{24} \xrightarrow{NH_3} HS(Bd)_nSH \xrightarrow{Sty} PS{-}b{-}PBd{-}b{-}PS \ .$$

3.1.2 Coupling of Monofunctional Telomers

Block copolymers can be produced from the coupling of ω-functional oligomers or telomers which exhibit complementary end-groups able to react together. The reactions most frequently used to achieve coupling between ω-functional polymers or cooliogmers are esterification, amidification and hydrosilylation. The difficulty may arise from the immiscibility between chemically unlike polymers [122]. The reaction medium initially tends to phase separate, reducing the probability of the antagonist functions reacting. However, once some block copolymer has been formed, it acts as a compatibilizer and the reaction medium becomes homogeneous. To avoid problems with immisicibility, low molecular weight polymers or telomers have to be chosen carefully.

According to the reactants, either diblock or triblock copolymers can be obtained. For instance, PEO-b-PDMS-b-PEO triblock copolymer and PEO-PDMS diblock copolymers were prepared in high yields by hydrosilylation of a telechelic PDMS which exhibits SiH functions (Mn = 1000) with monofunctional allyl-terminated PEO with Mn = 350 and 500 and telechelic diallyl PEO (Mn = 600), respectively [123]. Their dilute solution properties were investigated. Similarly, interesting PS-b-PDMS thermoplastics have been synthesized from a polystyrene fitted at chain end with a vinyl silane function which reacts with a PDMS bearing SiH end-groups [124]. In addition, hydrosilylation has been used to prepare original copolymers from α,ω-disilyl-PDMS 25 and either α,ω-diallyl-polysulfone [125] or α,ω-diallyl poly (L-lactide) (PLLA) as follows [126]:

$$H\left[OCHCO\atop {|\atop CH_3}\right]_n OC_2H_4O \left[COCHO\atop {|\atop CH_3}\right]_p H + H_2C=CHCH_2NCO \longrightarrow$$

PLLA

$$H_2C=CHCH_2NHCO_2-PLLA-OCONHCH_2CH=CH_2$$
26

$$25 + 26 \xrightarrow{H_2PtCl_6} \{PLLA-b-PDMS\}_x \cdot$$

The synthesis of poly(methacrylate)-b-poly(acrylate) cotelomers containing sequences of Mn = 620–6800 was recently performed in our laboratory according to the following scheme [127]:

$$CH_2=CH \atop {|\atop CO_2C_4H_9} + HSC_2H_4OH \longrightarrow HO-PBuA$$
27

$$MMA + HSC_2H_4OH \longrightarrow HO-PMMA$$
28

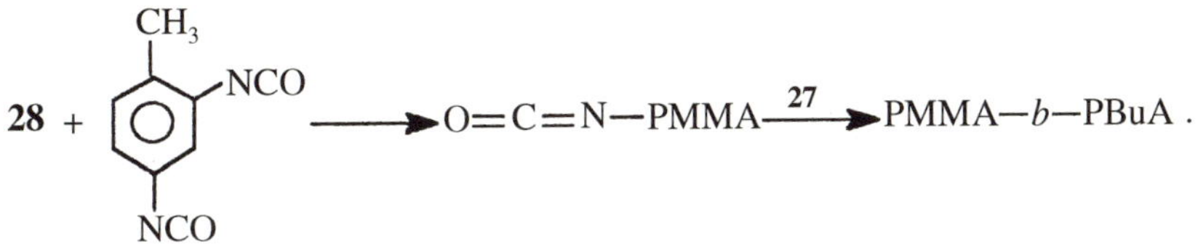

$$28 + \underset{NCO}{\underset{|}{\overset{CH_3}{\underset{|}{\bigcirc}}}}\!\!-NCO \longrightarrow O=C=N-PMMA \xrightarrow{27} PMMA-b-PBuA .$$

Similarly, the syntheses of polymethacrylate-b-poly(ethylene oxide), polymethacrylate-b-poly(methacrylic acid) and polymethacrylate-b-polysiloxane-b-polymethacrylate copolymers were achieved by end group transesterification of polymethacrylates prepared by radical polymerization in the presence of mercaptans [128].

In addition, Kennedy and Hongu [129] have prepared amphiphilic diblock cooligomers from the condensation of α-methoxy-ω-hydroxy polyoxyethylene with isocyanato-terminated polyisobutylene. In the same way, from α,ω-diisocyanate styrenic telomer **29**, a Japanese team [130] has prepared PFA-b-PS-b-PFA and PHEA-b-PS-b-PHEA triblock copolymers (where FA and HEA represents fluoro acrylate and 2-hydroxy ethyl acrylate, respectively), as follows:

$$OC=N-C_6H_4-S-S-C_6H_4-N=CO + Sty \longrightarrow O=C=N\text{ww}PS\text{ww}N=CO$$
$$\mathbf{29}$$

$$H-(CH-CH_2)_n-S-C_2H_4-R_2 + \mathbf{29} \longrightarrow -(CH-CH_2)_n-(Sty)_p-(CH_2-CH)_n-$$
$$\quad\quad | \quad\quad\quad\quad\quad\quad\quad\quad\quad\quad\quad\quad\quad\quad\quad\quad\quad\quad | \quad\quad\quad\quad\quad\quad\quad\quad\quad\quad\quad | $$
$$\quad\,CO_2R_1 \quad\quad\quad\quad\quad\quad\quad\quad\quad\quad\quad\quad\quad\quad\quad\quad CO_2R_1 \quad\quad\quad\quad\quad\quad\quad\quad CO_2R_1$$

$R_1 = C_4H_9$ or $C_2H_4C_4F_9$ or C_2H_4OH and $R_2 = OH$ or NH_2.

The original synthesis of compound **29** already described in previous work [131] deserves to be mentioned.

In addition, telomers which exhibit carboxy end-groups have been esterified with telechelic polyether diols and α,ω-difunctional PDMS have been mainly used as the central block, as in the three following examples [132–134]:

$$H(CHCH_2)_nSCH_2CO_2H + HO(CH_2)_3(SiO)_nSi(CH_2)_3OH \longrightarrow$$
$$\quad | $$
$$CO_2C_2H_4OH$$

PHEA—b—PDMS—b—PHEA

$$H(MMA)_nS(CH_2)_3Si(CH_3)(OEt)_2 + HO(SiO)_nH \longrightarrow$$

PMMA—b—PDMS—b—PMMA

$$H(CHCH_2)_nSiCl + Na^{\oplus\,\ominus}O(SiO)^{\ominus}\,Na^{\oplus} \longrightarrow PVAc—b—PDMS—b—PVAc\,.$$
$$\quad | $$
$$OAc$$

3.1.3 Conclusion

Functional telomers and oligomers are interesting precursors of block cooligomers. The monofunctional telomers lead to diblocks and coupling reactions require the use of telomers or oligomers with specific and antigonist functions. Triblock copolymers can be produced from telechelics. Traditional telomerization is still interesting because monofunctional telomers are very easy to obtain. On the other hand, bistelomerization may lead to lower yield because of either the less efficient transfer constant of the telomer/telogen or the numerous steps necessary to functionalize the telogen. However, new methods of reversible telomerization can be applied to obtain well-architectured molecules.

3.2 Controlled Radical Telomerization

Various telogens can be successfully used in order to obtain pseudo-living polymers. The oldest method is based on the cleavage of the C–I bond that has already led to industrial applications. However, more recently, the C–Cl bond has also been involved by the way of complexing ligand in redox telomerization. Other cleavages have also been investigated, e.g., C–C bond as in substituted tetraphenyl ethanes, C–S link in specific transfer agents, S–S bond for all the group of disulfides and also Si–Si bond.

3.2.1 C–Halogen Cleavage

3.2.1.1 C–Cl and C–Br Bonds

Recently, two American groups [135, 136] and a Japanese team [137] have demonstrated that controlled living radical polymerizations are possible from initiating systems that exhibit a C–Cl bond. Wang and Matyjaszewski [135, 138] and Wang et al. [139] show that the system composed of 1-phenyl ethyl chloride/CuX with X = Cl or Br, complexed by 2,2′-bipyridine is quite efficient. The CuX/bipyridine couple behaves as a halogen atom promotor, and offers a well-controlled polymerization of various monomers (styrene, methyl and butyl acrylates, MMA) with narrow molecular weight distributions (Mw/Mn = 1.14–1.50; Mn = 8500–120 000 and yield = 95%). Such an ATRP method (Atom Transfer Radical Polymerization) is successfully applied to the preparation of PS-b-PMA block copolymers (Mw/Mn = 1.35; Mn = 13 000 and yield = 95%) [135, 138]. Similarly, Percec and Barboin [136] show that the arenesulfonyl chloride/Cu(2,2′-bipyridine)$_n$Cl system also favors the living radical polymerization of styrene at 120 °C. According to the substituent on the arenesulfonyl group, Mn values and polydisperity indexes are in the 1500–4900 and 1.48–1.80 ranges, respectively.

In addition, Sawamoto et al. [137, 140] detail the radical living polymerization of MMA with an initiating system that consists of CCl_4, $RuCl_2(PPh_3)_3$ and methyl aluminium bis (2,6-ditertbutyl phenoxide) (MAO-type) with a constant polydispersity index vs conversion rate (Mw/Mn = 1.3). According to the authors, the polymerization is considered to proceed via repeating radical addition of MMA to a CCl$_4$-derived growing end having a C–Cl bond catalyzed by RuCl$_2$(PPh$_3$)$_3$ and MAO [137]. However, without any MAO, the monoadduct is usually produced whatever the monomer: styrene [141–143], MMA [141, 143], allyl acetate or non-conjugated dienes [144], vinylidene chloride [108, 145]. Furthermore, the radical polymerizations of MMA, styrene, vinyl acetate and isobutyl vinyl ether were investigated by Kameda and Ishii [146, 147] in the presence of chlorotriphenyl phosphine rhodium complexes. In the polymerization of MMA, the average molecular weight increases to 322 000 with an increase in monomer conversion.

Vinylidene chloride was previously investigated in the living redox telomerization using CCl$_4$ as the transfer agent, as follows [108, 145]:

$$CCl_4 + H_2C=CCl_2 \longrightarrow \underset{\mathbf{30}}{Cl_3CCH_2CCl_3}$$

$$\mathbf{30} \xrightarrow{H_2C=CCl_2} Cl_3CCH_2CCl_2CH_2CCl_3 \xrightarrow{H_2C=CCl_2} Cl_3C(CH_2CCl_2)_n{-}Cl \ .$$

However, the presence of the methylene groups between the polychlorinated groups decreases the reactivity of the telomer produced and such a telomerization slows down as soon as the triadduct is formed. Thus, it has been shown that this process could be activated by introducing electron-withdrawing atoms such as fluorine atoms adjacent to trichloromethyl end-groups [148]:

$$CCl_4 + (n+1)CF_2=CCl_2 \longrightarrow Cl_3C(CF_2CCl_2)_n{-}CF_2CCl_3 \ .$$

3.2.1.2 C–I Bond

The cleavage of C–I bond can be achieved by various methods [149–151]. However, from well-chosen monomers, two main ways have been developed in order to control telomerization from alkyl iodides: Iodine Transfer Polymerization (ITP) and degenerative transfer.

3.2.1.2.1 Iodine Transfer Polymerization

Iodine transfer polymerization was one of the radical living processes being developed in the late 1970s by Tatemoto et al. [150–156]. It originates from the radical telomerizations pioneered by Hanford and Joyce [157] and Haszeldine [158] in the 1940s as mentioned in various reviews on telomerization [159–162]. Actually, it is necessary to use (per) fluoroalkyl iodides because their highly electron-withdrawing (per) fluorinated group R_F produces the lowest level of the CF$_2$–I bond dissociation energy, such a C–I cleavage not being possible in $R_FCH_2CH_2I$. Various fluorinated monomers have been successfully used in ITP. Basic similarities in these living polymerization system are found in the stepwise growth of polymeric chains at each active species. The active living center, generally located at end-groups of the growing polymer, has the same reactivity at any time during polymerization even when the reaction is stopped [150–152]. In the case of ITP of fluoro olefin, the terminal active bond is always the C–I bond originating from the initial iodine-containing chain transfer agent and monomer as follows:

$$C_nF_{2n+1}-I + (p+1)H_2C=CF_2 \xrightarrow{R^{\cdot} \text{ or } \Delta} C_nF_{2n+1}-(C_2H_2F_2)_p-CH_2CF_2-I.$$

Usually molecular weights are not higher than 30 000 and yet polydispersity is narrow (1.2–1.3) [151, 163, 164].

Tatemoto et al. used peroxides as initiators of polymerization and described the mechanism by Scheme 2:

Initiation
$$R-O-O-R \xrightarrow{\Delta} RO^{\cdot} \tag{1}$$

$$RO^{\cdot} + R_FI \longrightarrow \text{products} + R_F^{\cdot} \tag{1'}$$

Propagation
$$R_F^{\cdot} + M \longrightarrow R_FM^{\cdot} \tag{2}$$

$$R_FM^{\cdot} + nM \longrightarrow R_F(M)_{n+1}^{\cdot} \tag{2'}$$

Transfer
$$R_F(M)_{n+1}^{\cdot} + R_FI \longrightarrow R_F(M)_{n+1}I + R_F^{\cdot} \tag{3}$$

$$R_F(M)_m^{\cdot} + R_F(M)_nI \longrightarrow R_F(M)_mI + R_FM_n^{\cdot} \quad (m>n) \tag{3'}$$

Termination
$$2R^{\cdot} \longrightarrow R-R \tag{4}$$

$$R_FM_n^{\cdot} + R_FM_m^{\cdot} \longrightarrow R_FM_{n+m}R_F \tag{4'}$$

Scheme 2. Mechanism of Iodine Transfer Polymerization

The transfer constants of each R_FI have to be taken into account in order to produce high or low molecular weight telomers. However, if standard telomerization steps, especially step (3') are taken into consideration, then living radical polymerization would become possible. Successive chain transfer reactions through step (3') are often observed. Therefore, such chain transfer reactions extending up to higher molecular weight ranges are key controlling factors to obtain satisfactory results in ITP. The situation would be more idealized by minimizing the amount of initiator in step (1) and would contribute to delay possibilities of steps (4) and (4'). Another improvement is also possible by the use of diiodide and polyiodide compounds [165].

Several investigations have shown that iodine transfer polymerization can be processable by emulsion or radical initiations. When emulsion initiation is chosen, a perfluoroalkyl iodide is involved and limits the molecular weights [166–168]. This will not be described here but several articles and patents from Tatemoto et al. can be suggested [150–156, 165], these using ammonium persulfate as the initiator and involving tetrafluoroethylene (TFE), vinylidene fluoride (VDF) and hexafluoropropene (HFP) as the monomers. Actually these fluoroelastomers produced by ITP can be peroxide-curable and lead to commerically available DAIEL [164, 168] produced by Daikin. Such a polymer is stable up to 200 °C and finds many high technology applications in areas such as transportation and electronics.

The most remarkable use of living polymerization is the preparation of block copolymers. The practical method for obtaining such copolymers is by direct cotelomerization of two fluoromonomers with an efficient transfer agent. For instance, Tatemoto and Morita [153] used a mixture of VDF/HFP in 46/54% molar ratio and IC_4F_8I in the presence of trichloroperfluorohexanoyl peroxide as the catalyst and obtained $I(HFP)_a(VDF)_bC_4F_8(VDF)_c(HFP)_dI$ that exhibited an average number molecular weight of 3300 with a polydispersity of 1.27. It consists of a stepwise cotelomerization as follows:

$$R_FI + nM_1 \longrightarrow R_F(M_1)_nI$$
$$\mathbf{31}$$

$$\mathbf{31} + mM_2 \longrightarrow R_F(M_1)_n(M_2)_mI$$

where M_1 and M_2 represent fluorinated monomers or a group of fluoroolefins which can be specially chosen to bring softness and hardness, respectively, leading to a thermoplastic elastomer [168–172]. For instance, a hard segment sequence can be composed of E/HFP/TFE in 43/8/49 molar ratio (E represents ethylene) whereas the soft part can consist of TFE/HFP/VDF in 20/30/50 molar ratio [171]. Similar schemes can also be applied to α,ω-diiodides [167]. Furthermore, the introduction of hard segments such as alternated copolymers of TFE and E, or PVDF, led to commercially available DAIEL Thermoplastic [164, 166, 172]. They exhibit very interesting properties such as high specific volume (1.90), high melting point (160–200 °C), high thermostability up to 380–400 °C, a refractive index of 1.357 and good surface properties ($\gamma_c \sim$ 19.6–20.5 dyne/cm). These characteristics offer excellent resistance against corrosive chemicals, strong acids, fuels, and oils. In addition, the tensile modulus is close of that of cured fluoroelastomers.

This concept was also exploited in order to prepare living and well defined tetrafluoroethylene telomers in which the telomer produced acts as a further telogen as follows:

$$C_2F_5I \xrightarrow{C_2F_4} C_4F_9I \xrightarrow{C_2F_4} C_6F_{13}I \xrightarrow{C_2F_4} C_8F_{17}I \longrightarrow C_nF_{2n+1}I.$$

Such a living telomerization can be initiated either thermally or in the presence of radical initiators or redox catalysts [173, 174]. The application of such a process allows the step-by-step syntheses of block copolymers either from fluoroalkyl iodides or α,ω-diiodoperfluoroalkanes. Investigations by Chambers et al. [175] or those performed in our laboratory have led to the extensive use of fluoroalkyl iodides with chlorotrifluoroethylene (CTFE) for the preparation of efficient transfer agents X–Y with X = Y = I [175] or Br [176] and also X = I and Y = Cl [177] or F [178, 179]:

$$X-Y + CF_2=CFCl \longrightarrow X(C_2F_3Cl)_nY$$

Chambers et al. obtained difunctional oligomers especially for brominated telogens [175]. It has been observed, however, that from ICl or (IF) generated in situ from iodine and iodine pentafluoride, monoadducts were obtained as the sole product [178, 179]. Addition of (IF) to CTFE yields CF_3CFClI which was successfully used as the telogen for the telomerizations of CTFE, hexafluoropropene (HFP) [180] or for the stepwise cotelomerizations of CTFE/HFP, CTFE/HFP/VDF, HFP/VDF and HFP/trifluoroethylene [181]. All these cotelomers were successfully end-capped by ethylene, allowing further functionalization [182]. Acutally the CFClI end-group shows a better reactivity than CF_2I. Similar telomerizations of VDF [183], trifluoroethylene [184], tetrafluoroethylene [173, 174], hexafluoropropene [185] and CTFE [179] have been achieved. All these monomers behave differently in both initiation and reactivity. Furthermore, the corresponding telomers exhibit specific defects of chaining and are produced with various degrees of polymerization. Interestingly, CTFE telomers which have CFClI end-groups are more reactive than perfluoroalkyl iodide telogens. In the case of the telomerization of CTFE, these last transfer agents lead to poor telogen conversion and higher molecular weight telomers than those produced from other fluorinated olefins (e.g., VDF, HFP, trifluoroethylene).

Furthermore, stepwise cotelomerizations of various commercially available fluoromonomers lead to interesting high fluorine-containing derivatives as follows:

$$C_4F_9I + CH_2=CF_2 \longrightarrow C_4F_9(C_2H_2F_2)_nI \xrightarrow{HFP} C_4F_9(VDF)_n(HFP)I \quad [183]$$

$$iC_3F_7I + CH_2=CF_2 \longrightarrow iC_3F_7(C_2H_2F_2)_nI \xrightarrow{HFP}$$

$$iC_3F_7(C_2H_2F_2)_nCF_2CFICF_3 \quad [181, 183, 186]$$

$$C_nF_{2n+1}I + HFP \longrightarrow C_nF_{2n+1}(HFP)I \xrightarrow{VDF} C_nF_{2n+1}(HFP)(VDF)_pI \quad [182, 185]$$
$n = 4, 6, 8$

$\downarrow C_2F_3H \qquad\qquad\qquad \downarrow C_2F_3H$

$C_nF_{2n+1}(HFP)(C_2F_3H)_xI \qquad F(TFE)_n(HFP)(VDF)_p(C_2H_3H)_mI$.

All these cotelomers were successfully end-capped with ethylene offering new ω-functionalized fluorocompounds [182]. In addition, α,ω-diiodofluoroalkanes allow access to well-defined block cotelomers. Tortelli and Tonelli [187] and Baum et al. [188, 189] carried out the synthesis of such telechelic products by heating iodine crystal with tetrafluoroethylene. Such halogenated reactants were successfully involved in telomerization of vinylidene fluoride (VDF) [190] or hexafluoropropene (HFP) [191] to form α,ω-diiodo VDF/TFE/VDF or α,ω-diiodo HFP/TFE/HFP triblock cotelomers. These products are potential starting materials for Viton-type multiblock cotelomers [192] as shown in the following examples:

$$I(VDF)_p(TFE)_n(VDF)_qI + HFP \longrightarrow I(HFP)_x(VDF)_p(TFE)_n(VDF)_q(HFP)_yI$$

$$I(HFP)(TFE)_n(HFP)_tI + VDF \longrightarrow I(VDF)_z(HFP)(TFE)_n(HFP)_t(VDF)_\omega I .$$

The novel α,ω-diiodides underwent functionalization, especially for the preparation of original fluorinated non-conjugated dienes [193] utilized in the preparation of hybrid fluorosilicones [194] that exhibit excellent properties at low and high temperatures.

3.2.1.2.2 Degenerative Transfer

Matyjaszewski et al. [195] introduced the concept of degenerative transfer, based on a thermodynamically neutral exchange reaction between active and dormant species. The main monomers studied are MMA, butyl acrylate and styrene. First, the living radical polymerization of MMA is investigated in the presence of methyl 2-iodo-2-methyl propionate produced from the addition of HI to MMA. Such a reaction involves the transfer of the iodine atom of this latter compound to the polymeric growing chain leading to an ω-iodo PMMA and a methyl-2-methyl propionate radical. Iodinated PMMA cannot react directly with MMA and it cannot react with another ω-iodo PMMA. This step is followed by the transfer of the iodine moiety to such a radical leading to the growing radical able to react with another MMA molecule and to some extent to behave in a living process yielding to polymers of controlled molecular weights, as follows:

$$(MMA)_n\overset{CH_3}{\underset{CO_2CH_3}{C^\bullet}} + CH_3-\overset{CH_3}{\underset{CO_2CH_3}{\overset{|}{C}}}-I \rightleftharpoons (MMA)_n-\overset{CH_3}{\underset{CO_2CH_3}{\overset{|}{C}}}-I + CH_3-\overset{CH_3}{\underset{CO_2CH_3}{C^\bullet}}$$

$$\quad\quad\quad 32 \quad\quad\quad\quad\quad 33 \quad\quad\quad\quad\quad\quad 34$$

$$34 \xrightarrow{32} (MMA)_p-\overset{CH_3}{\underset{CO_2CH_3}{C^\bullet}} \xrightarrow{MMA} (MMA)_{p+1}-\overset{CH_3}{\underset{CO_2CH_3}{C^\bullet}} .$$

However, in order to undergo a successful living radical polymerization, **33** needs to be a good transfer agent and the polymeric **34** derivative must exhibit a weak cleavage of the C–I bond. Such a process has been shown to be successful in the living radical polymerization of styrene from 100 °C and in the presence of 1-iodo-1-phenyl ethane generated by the addition of HI to styrene [195]. Thus, the use of alkyl iodides for degenerative transfer favors a well-controlled radical process leading to well-defined polymers [196]. For instance, this team describes the interesting stepwise block copolymerizations of butyl acrylate and styrene. The first method concerns the obtaining of poly(butylacrylate) followed by insertion of styrene base units. The experimental M_n and M_w/M_n values of the diblock copolymers produced ranged from 6 300 to 30 000 and 1.50 to 1.90, respectively. On the other hand, sequential radical copolymerization of butyl acrylate initiated with polystyrene at 70 °C led to diblock copolymers with experimental M_n and M_w/M_n values of 7 600 – 30 100 and 1.40 – 2.10, respectively [197].

3.2.2 C–C Cleavage

Bulky organocompounds which exhibit steric hindered carbon–carbon bonds have been used as initiators of radical polymerization. They can be either tetraphenylalkanes, reactants containing trityl end-groups or branched perfluoroalkanes. The first reaction was presented in 1967 by Borsig et al. [198] who studied the radical polymerization of MMA in the presence of 3,3,4,4-tetraphenyl hexane and 1,1,2,2-tetraphenyl cyclopentane. The kinetics of decomposition of both these initiators were investigated [199]. The latter initiator led to biradicals shown to yield high molecular weight polymers. Interestingly, these polymerizations exhibit a living character and polymerization rate constants were determined [198].

3.2.2.1 Substituted Tetraphenyl Ethanes

A series of at least 14 papers [200–208] have been published dealing with the synthesis of telechelic polymers or block copolymers from the radical polymerization of various vinyl monomers with substituted 1,1,2,2-tetraphenyl ethanes. These aromatic compounds, known for over a century [209], are efficient in radical polymerization [201, 210]. They behave as both initiators and terminating agents [200] that can be involved in living radical polymerization as illustrated in the following reaction:

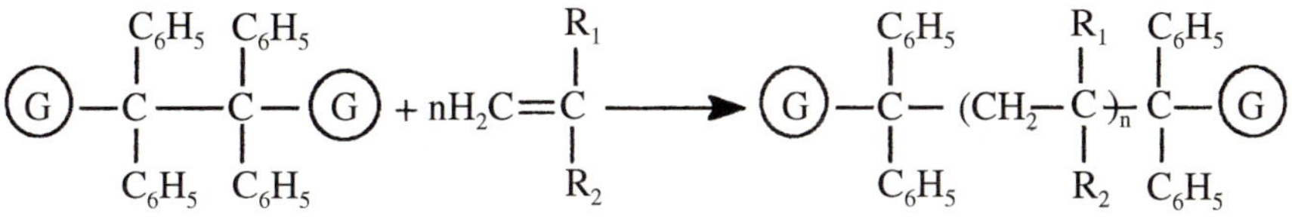

$$\text{(G)}-\underset{\underset{C_6H_5}{|}}{\overset{\overset{C_6H_5}{|}}{C}}-\underset{\underset{C_6H_5}{|}}{\overset{\overset{C_6H_5}{|}}{C}}-\text{(G)} + nH_2C=\underset{\underset{R_2}{|}}{\overset{\overset{R_1}{|}}{C}} \longrightarrow \text{(G)}-\underset{\underset{C_6H_5}{|}}{\overset{\overset{C_6H_5}{|}}{C}}-(CH_2-\underset{\underset{R_2}{|}}{\overset{\overset{R_1}{|}}{C}})_n-\underset{\underset{C_6H_5}{|}}{\overset{\overset{C_6H_5}{|}}{C}}-\text{(G)}$$

with $\boxed{G}$ = $OC_6H_5, CN, OSi(CH_3)_3$

R_1 = H or CH_3, $R_2 = C_6H_5$, or $R_1 = CH_3$ and $R_2 = CO_2R$

R = H, CH_3, nC_4H_9, tC_4H_9, $CH_2C_6H_5$.

These oligomers have shown novel applications, such as being efficient agents for thermal curing of unsaturated polyester resins [200]. Interestingly, methacrylate oligomers synthesized via this method are original macroinitiators for further polymerizations that can be performed at mild temperatures leading to diblock copolymers [203, 204]. These initiators exhibit a labile terminal C–C bond between the last monomeric unit and the initiator-derived radical, cleaving at 80 °C (in the case of copolymerization of styrene) and resulting in faster progress of the polymerization than that involving 1,2-diphenoxy-1,1,2,2-tetraphenyl ethane as the initiator, as follows:

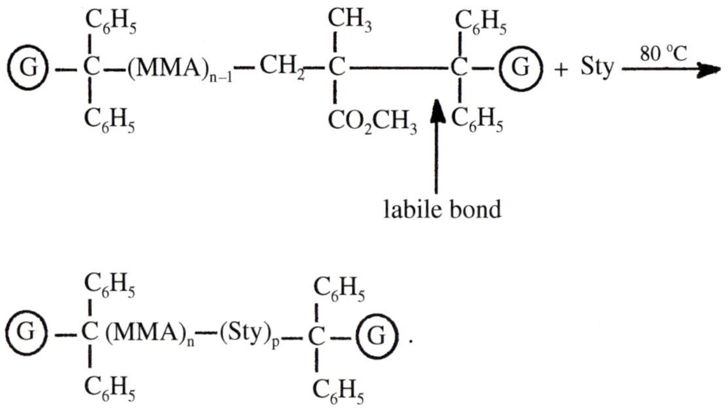

$$\boxed{G} - \underset{C_6H_5}{\overset{C_6H_5}{C}} - (MMA)_{n-1} - CH_2 - \underset{CO_2CH_3}{\overset{CH_3}{C}} - \underset{C_6H_5}{\overset{C_6H_5}{C}} - \boxed{G} + Sty \xrightarrow{80\,°C}$$

labile bond

$$\boxed{G} - \underset{C_6H_5}{\overset{C_6H_5}{C}}(MMA)_n - (Sty)_p - \underset{C_6H_5}{\overset{C_6H_5}{C}} - \boxed{G} \,.$$

From this method, telechelic and amphiphilic diblock copolymers containing hydrophilic methacrylic acid sequences and hydrophobic styrene segments were obtained [205].

Such process is described as "resuscitable free radical" polymerization [204]. It has been applied to the synthesis of poly(α-Me styrene)-b-poly(methyl methacrylate) copolymers from α-methyl styrene oligomers [206] and of amphiphilic copolymers containing polymethacrylic acid and PS sequences [205]. However, macroinitiation from styrene oligomers failed as a probable consequence of the too high stability of the end-groups.

Braun [200] reported that oligobenzopinacoles obtained by UV irradiation of bisbenzophenones and bishydroles are efficient initiators for the polymerization of styrene at 70–100 °C, allowing the formation of the following diblock copolymer:

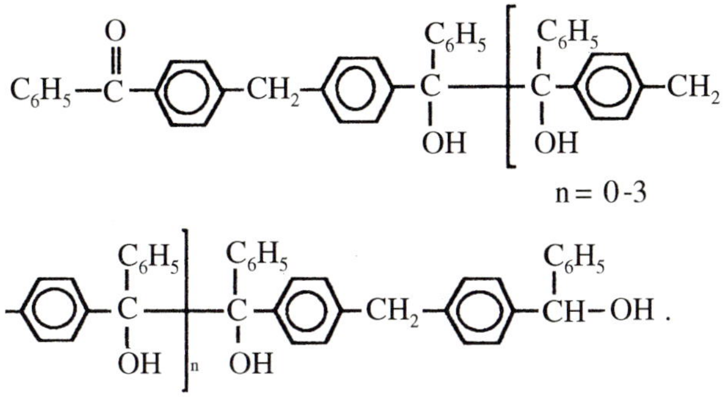

$$C_6H_5-\overset{O}{\underset{}{C}}-\underset{}{\bigcirc}-CH_2-\underset{}{\bigcirc}-\underset{OH}{\overset{C_6H_5}{\underset{|}{C}}}-\left[\underset{OH}{\overset{C_6H_5}{\underset{|}{C}}}-\underset{}{\bigcirc}-CH_2\right.$$

$$\left.-\underset{}{\bigcirc}-\underset{\underset{OH}{|}}{\overset{C_6H_5}{\underset{|}{C}}}\right]_n-\underset{\underset{OH}{|}}{\overset{C_6H_5}{\underset{|}{C}}}-\underset{}{\bigcirc}-CH_2-\underset{}{\bigcirc}-\underset{\underset{}{|}}{\overset{C_6H_5}{\underset{|}{CH}}}-OH.$$

n = 0-3

According to a similar process, macromolecular bis(silyl benzopinacolate) derivatives have been of particular interest in the synthesis of multiblock copolymers containing siloxane sequences proposed by Crivello et al. [211]. These authors used a high molecular weight PDMS macroinitiator **35**.

Compound **35** contains a thermolabile C–C bond which, on thermally induced fragmentation, yields a high proportion of macroradicals. The PDMS diradical also acts as a counter radical and can undergo chain extension at both ends in the presence of vinylic monomers (acrylonitrile, maleic anhydride, diethylfumarate) or styrenic monomers, leading to diblock copolymers in 95% yield according to the following scheme [211, 212]:

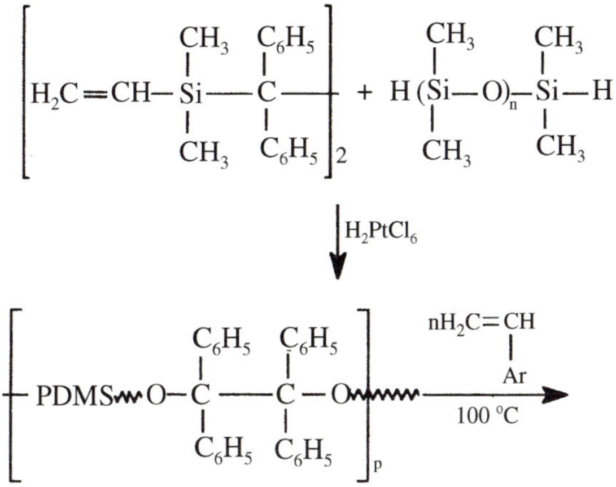

$$\left[H_2C=CH-\underset{CH_3}{\overset{CH_3}{\underset{|}{Si}}}-\underset{C_6H_5}{\overset{C_6H_5}{\underset{|}{C}}}-\right]_2 + H\underset{CH_3}{\overset{CH_3}{\underset{|}{(Si}}}-O)_n\underset{CH_3}{\overset{CH_3}{\underset{|}{Si}}}-H$$

$$\downarrow H_2PtCl_6$$

$$\left[-PDMS\sim O-\underset{C_6H_5}{\overset{C_6H_5}{\underset{|}{C}}}-\underset{C_6H_5}{\overset{C_6H_5}{\underset{|}{C}}}-O\sim\right]_p \xrightarrow[100\,°C]{nH_2C=CH-Ar}$$

35

$$\left[-PDMS\sim O-\underset{C_6H_5}{\overset{C_6H_5}{\underset{|}{C}}}-(CH_2-\underset{Ar}{\overset{}{\underset{|}{CH}}})_n-\right]_p$$

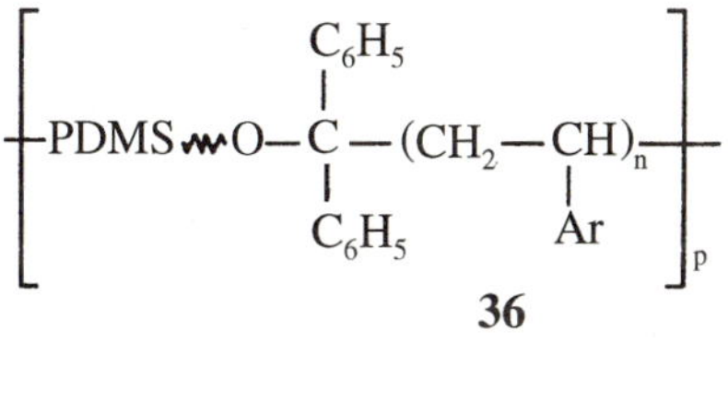

36

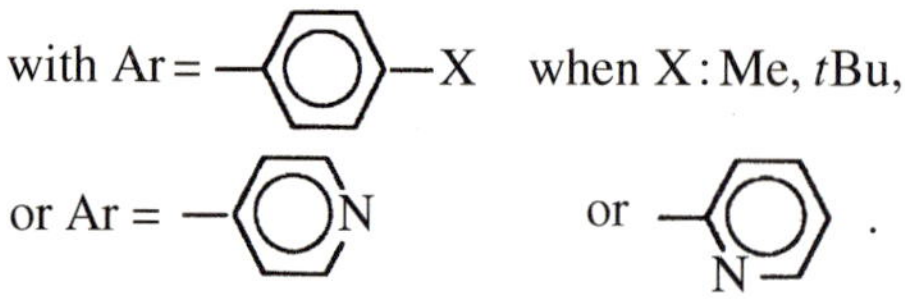

with Ar = –⟨C₆H₄⟩–X   when X: Me, tBu, OMe, Cl, Br

or Ar = –⟨C₅H₃N⟩   or   –⟨C₅H₄N⟩ .

Extensive investigations on these multiblock copolymers were performed for both synthesis and properties. Their synthesis includes detailed characterizations, living radical behavior, mechanism and the obtaining of high molecular weight copolymers with repeat sequences larger than 5. Their mechanical and thermal properties (Tg > 130 °C) allow them to be used as high temperature thermoplastic elastomers and some of them were thermally more stable than polystyrene. Interestingly, PDMS-b-PS prepolymers could undergo a further thermolysis in the presence of methyl methacrylate in order to yield multitriblock copolymers –[PDMS-b-PS-b-PMMA]$_n$– [212] as follows:

$$36 \xrightarrow[85\,°C/3\,h]{pMMA} \left[PDMS-O-\underset{C_6H_5}{\overset{C_6H_5}{C}}-(Sty)_n-(MMA)_p \right]_q .$$

Similar isomeric initiator as those used by Bledzki and Braun have also been used successfully in radical living polymerization of vinyl monomers for obtaining diblock copolymers.

3.2.2.2 Trityl Radical

The triphenyl methyl or trityl radical behaves as a radical trap and favors the polymerization-termination which is thermoreversible and thus allows the insertion of a new polymeric sequence. In 1982, Otsu et al. [49, 213, 214] proposed an interesting example involving phenylazotriphenylmethane as Initer (initiator-terminator) able to initiate a free radical polymerization from the phenyl radical. Alternatively, the trityl end-capped polymer can be utilized as an original macroiniter for the polymerization of a second monomer and yields block copolymers as follows:

$$C_6H_5-N=N-C(C_6H_5)_3 \xrightarrow{\Delta} C_6H_5^{\bullet} + (C_6H_5)_3C^{\bullet} + N_2$$

$$C_6H_5^{\bullet} + M_1 \longrightarrow C_6H_5M_1^{\bullet} \xrightarrow{(n-1)M_1} C_6H_5(M_1)_n^{\bullet}$$

$$C_6H_5(M_1)_n^{\bullet} + \dot{C}(C_6H_5)_3 \underset{\Delta}{\rightleftarrows} C_6H_5(M_1)_n C(C_6H_5)_3$$
$$37$$

$$37 \xrightarrow{mM_2} C_6H_5-(M_1)_n-(M_2)_m-C(C_6H_5)_3 .$$

Various methacrylic-styrene copolymers were prepared in which the reactivity of methacrylate monomers used in the first step decreases in the order: MMA > BuMA > benzyl methacrylate. For instance, the bulk polymerization of MMA with such an aromatic azo compound proceeds via a living radical mechanism and the sterically crowded $C-C(C_6H_5)_3$ terminal bond of polymethacrylate **37** can be cleaved thermally to produce α,ω-diaromatic PMMA-*b*-PS block copolymers in 48–72% yield.

3.2.2.3 Hindered Branched Perfluoroalkanes

The process has been extended to fluorinated radical initiators, their highly electron-withdrawing character favoring the homolytic C–C cleavage. Hexafluoropropene trimers reacted with fluorine to give hindered branched-chain perfluoroalkanes able to be easily cleaved at a temperature lower than 200 °C leading to perfluorinated radicals. These radicals can initiate the polymerization of vinyl and especially fluorinated monomers (e.g., tetrafluoroethylene, chlorotrifluoroethylene, vinylidene fluoride and tetrafluoroethylene/hexafluoropropene) [215, 216]. The evidence was shown by monitoring the thermal decomposition of the perfluorocompounds in the presence of molecular halogens [217] leading to branched perfluoroalkyl halides:

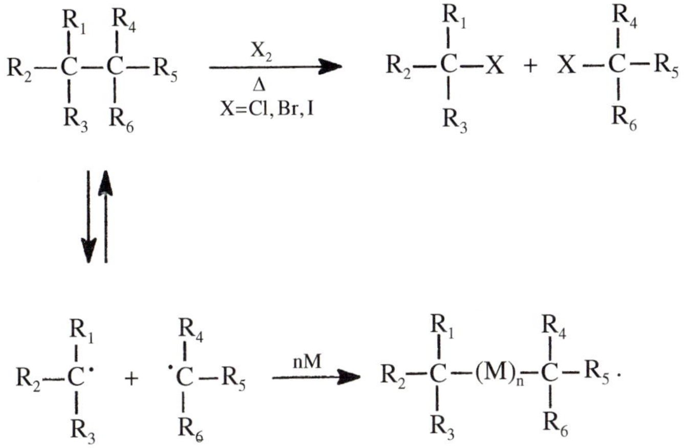

$$R_2-\underset{\underset{R_3}{|}}{\overset{\overset{R_1}{|}}{C}}-\underset{\underset{R_6}{|}}{\overset{\overset{R_4}{|}}{C}}-R_5 \xrightarrow[\Delta]{X_2} R_2-\underset{\underset{R_3}{|}}{\overset{\overset{R_1}{|}}{C}}-X + X-\underset{\underset{R_6}{|}}{\overset{\overset{R_4}{|}}{C}}-R_5$$

X=Cl, Br, I

$$\updownarrow$$

$$R_2-\underset{\underset{R_3}{|}}{\overset{\overset{R_1}{|}}{C}}{}^\bullet + {}^\bullet\underset{\underset{R_6}{|}}{\overset{\overset{R_4}{|}}{C}}-R_5 \xrightarrow{nM} R_2-\underset{\underset{R_3}{|}}{\overset{\overset{R_1}{|}}{C}}-(M)_n-\underset{\underset{R_6}{|}}{\overset{\overset{R_4}{|}}{C}}-R_5 \ .$$

In this scheme, R_i represents a fluorine atom or a perfluorinated group and M a fluorinated monomer. However, such a method has not been extended to the synthesis of block copolymers.

3.2.3 C–S Cleavage

Few references can be found in the literature about telogens able to generate cleavable C–S bonds [49, 218, 219]. They have been successfully used in telomerization to produce living block cotelomers. Two main telogens have been proposed, diaromatic azo **38** [49] and trityl mercaptan (or triphenyl methyl mercaptan) **41** [218, 219] compounds. With **38**, cotelomers are obtained as follows:

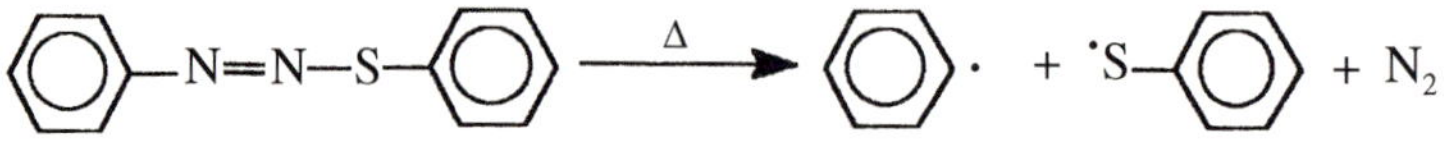

$$Ph-N=N-S-Ph \xrightarrow{\Delta} Ph\cdot + \cdot S-Ph + N_2$$
$$\mathbf{38} \qquad\qquad\qquad \mathbf{39}$$

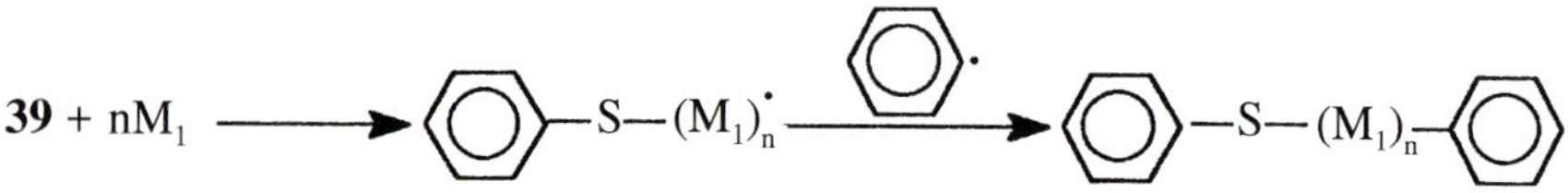

$$\mathbf{39} + nM_1 \longrightarrow Ph-S-(M_1)_n^{\cdot} \xrightarrow{Ph\cdot} Ph-S-(M_1)_n-Ph$$
$$\mathbf{40}$$

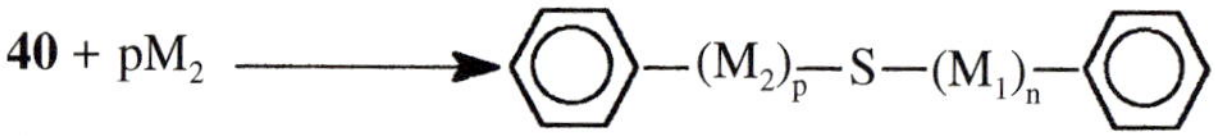

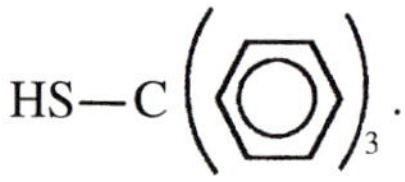

$$\mathbf{40} + pM_2 \longrightarrow Ph-(M_2)_p-S-(M_1)_n-Ph$$

$$HS-C(Ph)_3$$
$$\mathbf{41}$$

With compound **41**, stepwise cotelomerization of styrene (M_1) and MMA (M_2), in the presence of AIBN at 70 °C, led to the following diblock cotelomers:

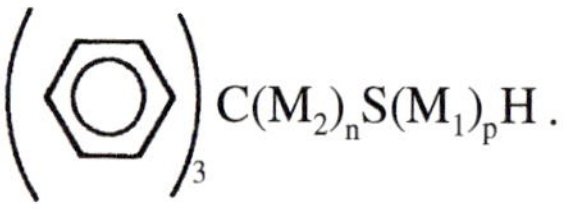

$$(Ph)_3 C(M_2)_n S(M_1)_p H.$$

The ω-trityl polymers obtained in the first step (Mn = 50 000–100 000) were able to initiate polymerization of MMA and styrene. Transfer constants of the aromatic mercaptan were determined (17.8 for styrene and 0.7 for MMA). In the presence of a second monomer, the authors noted a high increase of molecular weights and they were able to separate block copolymers from homopolymers. The molecular weights of the copolymers obtained and their compositions are given in Table 2 [218, 219].

The assumed structure of PS-*b*-PS is

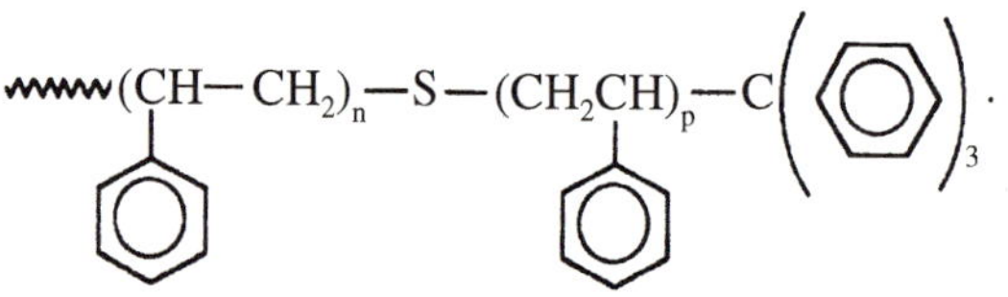

$$\sim\sim\sim(CH-CH_2)_n-S-(CH_2CH)_p-C(Ph)_3$$
with Ph groups on the CH and CH positions.

Such work is very interesting conceptually but still remains incomplete and requires further investigations in order to confirm the structures, and to check

Table 2. Percentages of homopolymers and block copolymers and molecular weights of PS-b-PMMA and PMMA-b-PS copolymers obtained by Yagci et al. [218, 219]

Macroinitiator	Homo PS (%)	Homo PMMA (%)	Block copolymer (%)
ω-$(C_6H_5)_3$C-PS	22	26	52 (110,000)
ω-$(C_6H_5)_3$C-PMMA	22	16	62 (140,000)
ω-$(C_6H_5)_3$C-PMMA	15	19	66 (145,000)

the nature of the cleavage occurring during the copolymerization. Furthermore, the authors do not explain the formation and the high amount of homopolymers probably due to the low transfer constant of trityl group.

3.2.4 S–S Cleavage

3.2.4.1 Introduction

Disulfide compounds including xanthogens and thiurams constitute an interesting group of versatile products which has led to many applications in macromolecular synthesis. Their recent development in radical polymerization was promoted by Otsu et al. [50, 213] for the synthesis of telechelic or α, ω-difunctional oligomers, even if Ferrington and Tobolsky [220] had already used them first in 1955. Otsu introduced the concept of "Iniferter" polymerization in calling the disulfide compounds used "Iniferters" because they are able to act as initiators, transfer agents and terminators. The Iniferter system differs from the "Infer" system developed by Kennedy [221] and applied to functionalizations of cationic polymers at chain-ends, since the radical issued from the homolytic decomposition of the initiator also acts as terminator.

A wide variety of disulfide derivatives has been proposed and among them the most simple R–S–S–R are those mainly used. Many investigations are still being conducted to increase the reactivity and the ease of cleavable of the S–S bond [12, 21]. Specific initiators include alcoyle disulfides **42**, where R represents an aromatic group [49, 222], dithiobenzoyl disulfides **43** [223], dialkyl xanthogen disulfides **44** [122], and finally thiuram disulfides **45** [224]:

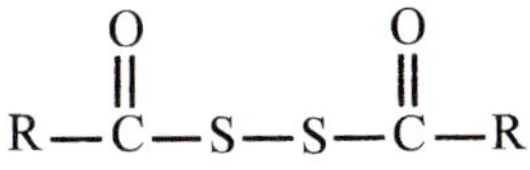

$$R-\underset{\underset{42}{}}{\overset{O}{\overset{\|}{C}}}-S-S-\overset{O}{\overset{\|}{C}}-R$$

$$R-\underset{\underset{43}{}}{\overset{S}{\overset{\|}{C}}}-S-S-\overset{S}{\overset{\|}{C}}-R$$

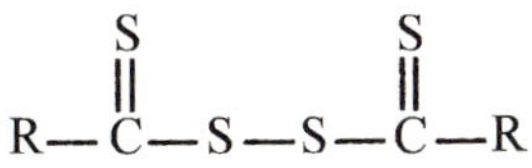

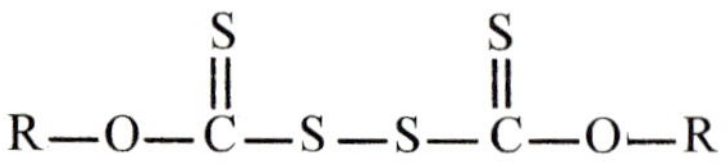

$$\underset{\textbf{44}}{R-O-\overset{\overset{S}{\|}}{C}-S-S-\overset{\overset{S}{\|}}{C}-O-R}$$

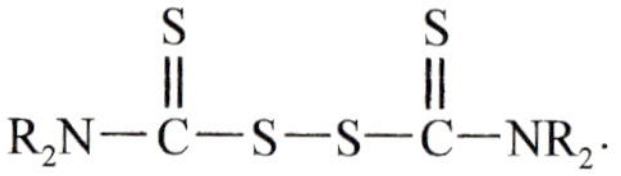

$$\underset{\textbf{45}}{R_2N-\overset{\overset{S}{\|}}{C}-S-S-\overset{\overset{S}{\|}}{C}-NR_2}.$$

If X_2 and M are the iniferter and the monomer, respectively, the reaction scheme involves five main reactions:

Initiation $\quad X_2 \xrightarrow[\text{or } h\nu]{\Delta} 2X^\bullet$

Propagation $\quad X^\bullet + M \longrightarrow XM^\bullet$

$\quad\quad\quad\quad\quad\quad XM^\bullet \xrightarrow{nM} XM^\bullet_{n+1}$

Transfer $\quad XM^\bullet_{n+1} + X_2 \xrightarrow{k_{tr}} X(M)_{n+1} X + X^\bullet$

Termination $\quad XM^\bullet_{n+1} + {}^\bullet X \underset{}{\overset{k_{te}}{\rightleftarrows}} X(M)_{n+1} X.$

This scheme shows three important differences compared with the classical reaction applied to the telogen formation already described [159–162]. First, X_2 compounds are able to initiate the polymerization readily, unlike telogens in telomerization. Second, X_2 derivatives can be symmetrical which is not the case for telogens (mercaptans, carbon tetrachloride, ω-CCl_3 compounds). Finally, and more important, in the Iniferter process the termination reaction can be reversible, allowing the insertion of new monomeric units. In other words, from the precursor macroiniferter $X(M)_{n+1}X$, block copolymer synthesis is possible. The synthesis and applications of macroiniferters for producing telechelics and block copolymers have been reviewed recently in details by Nair et al. [12], Otsu and Matsumoto [213] and Kumar et al. [225]. In this context, only the main concepts and models are summarized in the following section with, however, special attention being paid to the pseudo living character of the system.

3.2.4.2 Simple Disulfides

Since Otsu's investigations [222] showing that aromatic disulfides or tetrasulfides such as **46** and **47** are the most reactive owing to their high transfer constant [21] a variety of functionalized and substituted aromatic disulfides has been synthesized and applied to the preparation of telechelics:

Ph—(S)$_n$—Ph   and   Ph—CH$_2$S—SCH$_2$—Ph .

n = 2–4
46   **47**

Most pertinent examples are mentioned in the review by Clouet et al. [12] and various monomers were utilized as macroprecursors: styrene, MMA, alkyl acrylates, and isoprene. The synthesis of triblock copolymers can be easily achieved by condensation of these macroprecursors with monofunctional polymeric derivatives as shown in the following characteristic example [226]:

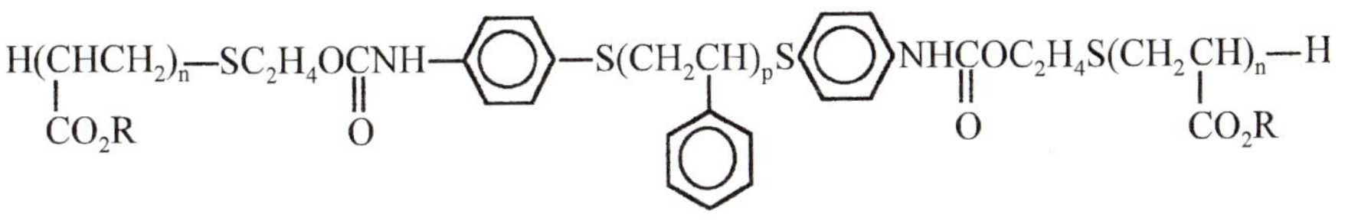

H(CHCH$_2$)$_n$—SC$_2$H$_4$OCNH—Ph—S(CH$_2$CH)$_p$S—Ph—NHCOC$_2$H$_4$S(CH$_2$CH)$_n$—H
  | || | || |
  CO$_2$R   O   Ph   O   CO$_2$R

with R = C$_2$H$_4$C$_4$F$_9$.

Although in this example the authors claimed no living character to the synthesis, Opresnik et al. [227, 228] described a similar synthesis in which some living character is seen. They also used disulfides as reversible termination agents in the presence of styrene, MMA and ethyl acrylate (EA). The first step involves the synthesis of polymeric precursor **48** under UV cleavage:

R—S—S—R + nM$_1$ $\xrightarrow{h\nu}$ RS(M$_1$)$_n$SR .

48

This Slovenic team has demonstrated that the best disulfides are aromatic bearing electronegative substituents with a positive resonance effect (NH$_2$ or Cl groups). Yields and molecular weights obtained increase with temperature and depend on the monomer structure with a decrease in the order EA > MMA > Sty. From these macroinitiators, and after purification by precipitation, the authors performed either chain extension with the same monomer or the synthesis of block copolymers. Interestingly, the living character is clearly shown by the increase of Mn value with both the monomer conversion and time. Furthermore, the amounts of homopolymer formed simultaneously are very low. This denotes that the generated RS° radical acts as a counter radical and reacts with the growing macroradical (almost exclusively) rather than with the second monomer.

Such a straightforward and promising investigation needs to be continued since it is an original way to synthesize block copolymers particularly well adapted to acrylic monomers.

3.2.4.3 Xanthogen Disulfides

Xanthogen disulfides, also called thio formic acid dithio OO' dialkylester, exhibit the following general structure:

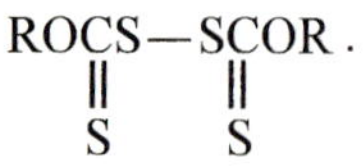

$$\underset{S}{ROC}\!\!-\!\!\underset{S}{SCOR}\,.$$
(where each C=S)

They are prepared from the addition of CS_2 to the corresponding alcohols in basic medium. They are known to exhibit high transfer constants ($C_T = 1\text{--}20$) [21]. Constanza et al. [210], Otsu and Yoshida [49], Uraneck [121] and Fokina et al. [229] intensively studied the polymerization of styrene, MMA and butadiene with these compounds in order to obtain telechelics with functional xanthogens. Furthermore, chemical change (e.g., hydrolysis) was also performed to achieve the synthesis of corresponding α,ω-dithiols [229] such as from **49**:

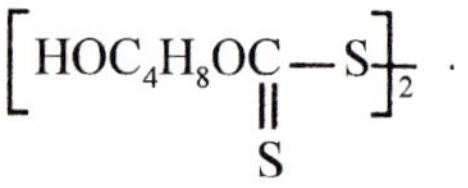

$$\left[HOC_4H_8O\underset{S}{\overset{\|}{C}}-S\right]_2.$$

49

One of the most interesting investigations on the stepwise polymerization was performed by Niwa et al. [230, 231] who prepared different diblock copolymers with, very often, a high increase of molecular weight and low polydispersity indexes. As an example of precursor of diblock copolymers, they prepared telechelic **50** as follows:

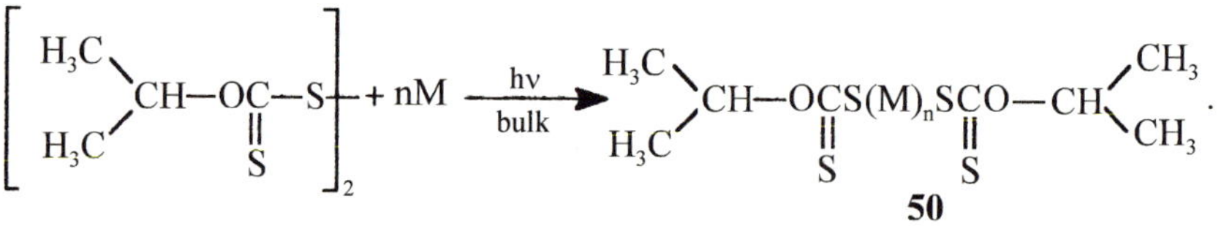

$$\left[\underset{H_3C}{\overset{H_3C}{\diagdown}}CH-O\underset{S}{\overset{\|}{C}}-S\right]_2 + nM \xrightarrow[\text{bulk}]{h\nu} \underset{H_3C}{\overset{H_3C}{\diagdown}}CH-O\underset{S}{\overset{\|}{C}}S(M)_n S\underset{S}{\overset{\|}{C}}O-CH\underset{CH_3}{\overset{CH_3}{\diagup}}.$$

50

In this synthesis, the reactivity of monomers M decreases in the classic order: methyl acrylate > MMA ≫ acrylonitrile ~ vinyl acetate > Sty, and DPn values increase linearly with the monomer conversion. Moreover, molecular weights range from 9000 (styrene) to 70 000 (methyl acrylate) and the yields are quite high after only 3 h (ca. 90%) [230, 231]. In addition, the authors observed that the initial value of DPn decreases with increase in the concentration of xanthogen disulfide [TX], an observation in agreement with the following kinetic equation:

$$\frac{1}{\overline{DP}_n} = \frac{A}{[M]} + C_{TX}\frac{[TX]}{[M]}.$$

In this equation, the second term only represents the normal behavior of the telomerization and the transfer constant of xanthogen disulfide (C_T), 4.44 and 0.29 for styrene and MMA, respectively. Such values are close to those previously determined for mercaptans [160–162].

It should be noted that the authors surprisingly described the kinetics of these reactions as a normal process and not as a living radical system.

3.2.4.4 Thiuram Disulfides

Thiuram disulfides are undoubtedly the most studied products. These molecules can be symmetrical or not, and mono- or multifunctional (especially thiourea groups) as shown in the following examples [232]:

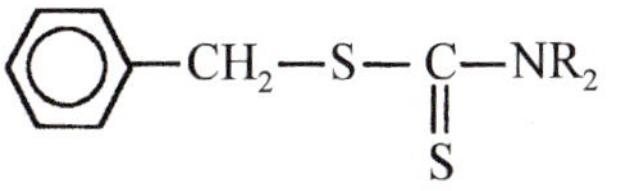

$$\text{Ph-CH}_2\text{-S-C(=S)-NR}_2 \quad ; \quad \begin{array}{c} R \\ R' \end{array}\!\!\!\text{N-C(=S)-(S)}_n\text{-C(=S)-N}\!\!\!\begin{array}{c} R \\ R' \end{array}$$

$$n = 2, 4$$

51 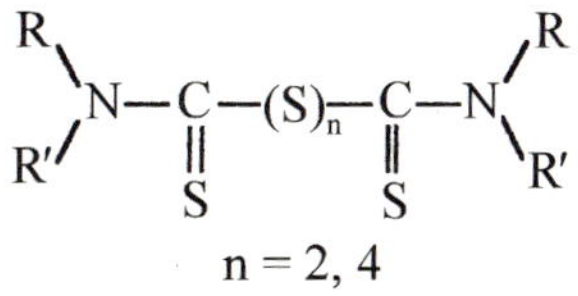   **52**

$$\text{1,2,4,5-C}_6\text{H}_2[\text{CH}_2\text{-S-C(=S)-NR}_2]_4$$

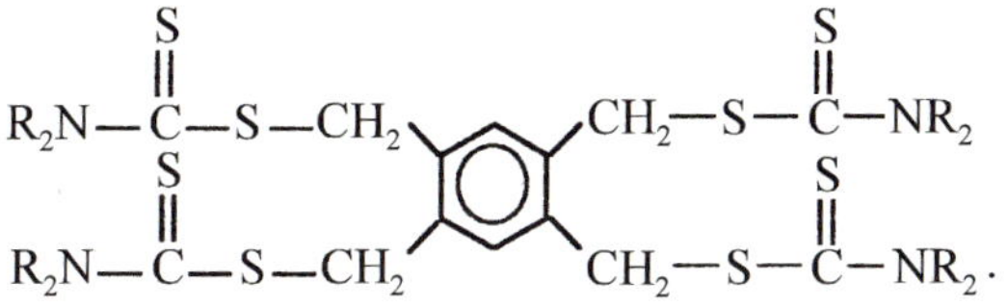

53

In addition, different macromolecular thiuram disulfides have been synthesized by Clouet et al. [12, 233] such as

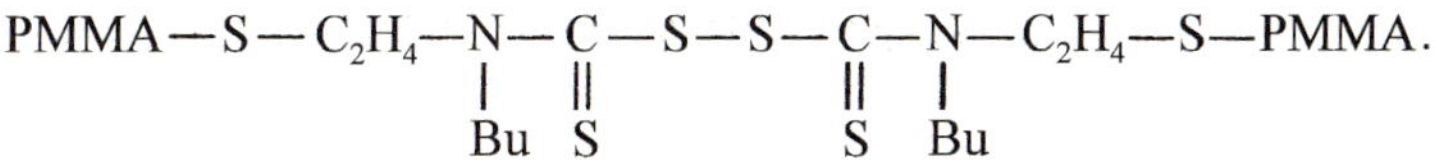

$$\text{PMMA-S-C}_2\text{H}_4\text{-N(Bu)-C(=S)-S-S-C(=S)-N(Bu)-C}_2\text{H}_4\text{-S-PMMA}.$$

From such compounds, series of di-, tri- and multiblock copolymers were prepared with a large variety of structures including soft, hard, hydrophobic, hydrophilic, silicone or phosphonamide sequences [12].

Investigations were mainly devoted to the synthesis of telechelic polymers and copolymers rather than to living radical polymerization. In particular, from 1960, Imoto et al. [234] started surveys on the synthesis of block copolymers from this method. Thus, polystyrene-b-poly(vinyl alcohol) diblock copolymer was obtained as follows:

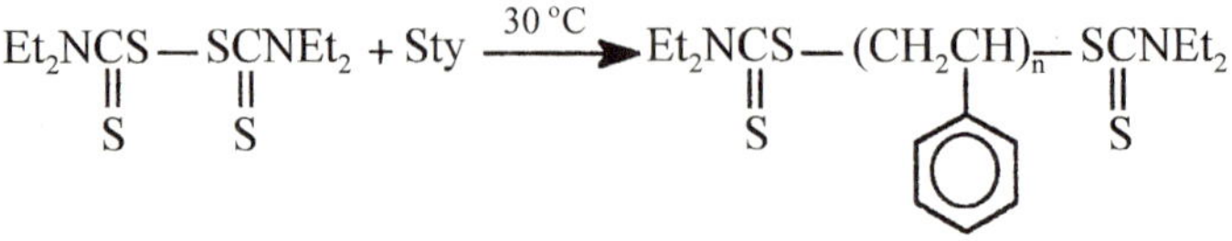

$$Et_2NCS-SCNEt_2 + Sty \xrightarrow{30\,°C} Et_2NCS-(CH_2CH)_{\overline{n}}-SCNEt_2$$
(with C=S groups and phenyl ring on CH)

54

$$54 + H_2C=CH-O\overset{O}{\overset{\|}{C}}CH_3 \xrightarrow[\text{ii) NaOH}]{\text{i) h}\nu} PVA-b-PS-b-PVA.$$
(PVA: polyvinylalcohol)

Actually, at that time, the yields were low and the syntheses were complex.

More than twenty years later, several teams have launched research programs to analyze and understand better the mechanism of the reactions involved. In particular, the demonstration of the living character of these polymerizations seems quite important. However, the analysis of the results shows that the behavior of thiurams or similar compounds strongly depends on the nature of the monomers. In the following, the cases of the two most common monomer types, styrene and acrylic monomers, will be detailed separately.

3.2.4.4.1 Styrene

Four main studies have been published on styrene in which the authors show that thiuram disulfides play the role of both initiator and counter radical. Otsu et al. [235] demonstrated with **51** and **52** ($n = 2$) compounds and model molecule **55** that the polymerization of styrene led to the formation of mono- and dithiuram telechelic PS with a behavior close to that of living radical polymerization:

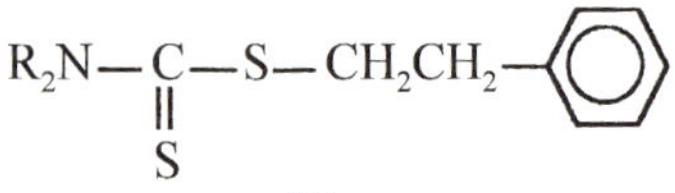

$$R_2N-\overset{\|}{\underset{S}{C}}-S-CH_2CH_2-\text{Ph}$$

55

The authors confirmed by ^{13}C NMR the presence of two terminal thiourea group:

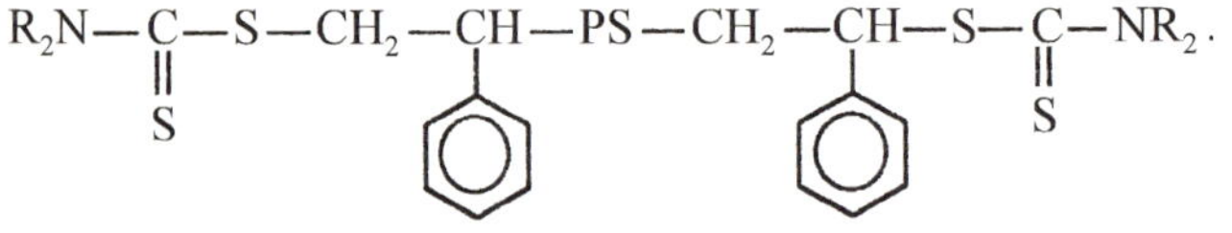

$$R_2N-\overset{\|}{\underset{S}{C}}-S-CH_2-CH(Ph)-PS-CH_2-CH(Ph)-S-\overset{\|}{\underset{S}{C}}-NR_2.$$

Furthermore, Turner and Blevins [236] and Van Kerckhoven et al. [237] observed with dithiocarbamates that, if the photochemically induced telomerization is monitored by a living character, the thermal reaction does not exhibit a living behavior and that the number average molecular weight remains practically constant during the polymerization. They show that the end-group **56** is stable up to 100 °C in contrast to the starting thiuram disulfide, but under

UV radiation, dissociation may occur:

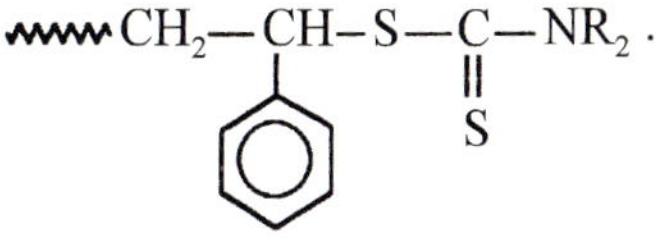

$$\text{\textasciitilde}CH_2-CH(C_6H_5)-S-C(=S)-NR_2$$

56

To complete these studies, Van Kerckhoven et al. [237] have investigated the behavior of styrene according to two key experiments: first, they showed by ^{13}C NMR that when an excess of thiuram disulfide is present, a dicapped structure is effectively obtained:

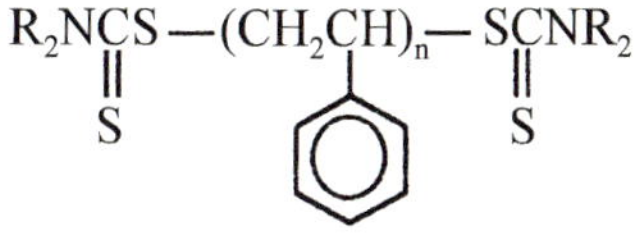

$$R_2N-C(=S)-S-(CH_2CH(C_6H_5))_n-S-C(=S)-NR_2$$

and, second, using an excess of AIBN as the initiator, they demonstrated the formation of the dual compound

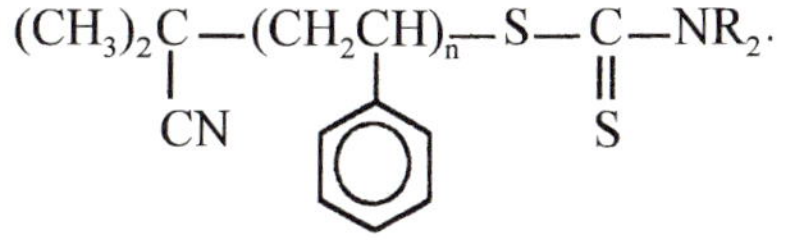

$$(CH_3)_2C(CN)-(CH_2CH(C_6H_5))_n-S-C(=S)-NR_2$$

This clearly shows that it is necessary to separate the initiating and terminating groups except when a symmetrical thiuram is used as the sole reactant.

In conclusion, styrene can be considered as an "ideal" monomer in photochemically induced polymerization and can be used successfully in quasi living polymerizations.

3.2.4.4.2 Acrylic Monomers

All authors seem to claim that for MMA the photochemical chain extension of thiocarbamate-terminated PMMA appears to be a more complicated process. Turner and Blevins [236] have shown that, during the process, the formation of CS_2 is not negligible, unlike the polymerization of styrene, and they explained its production by the following scheme:

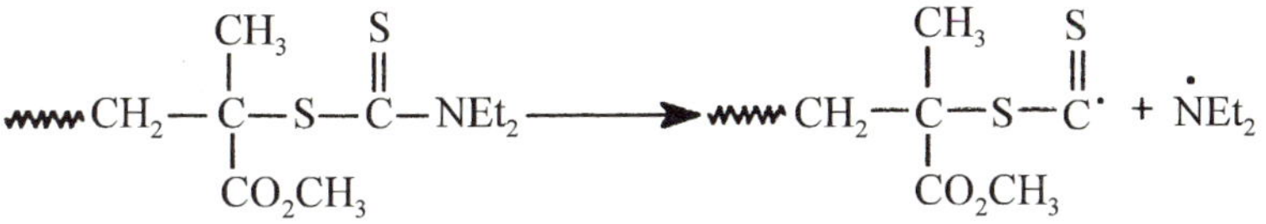

$$\text{\textasciitilde}CH_2-C(CH_3)(CO_2CH_3)-S-C(=S)-NEt_2 \longrightarrow \text{\textasciitilde}CH_2-C(CH_3)(CO_2CH_3)-S-\dot{C}(=S) + \dot{N}Et_2$$

57

$$\mathbf{57} \longrightarrow \text{\textasciitilde}CH_2-\dot{C}(CH_3)(CO_2CH_3) + CS_2 + \dot{N}Et_2$$

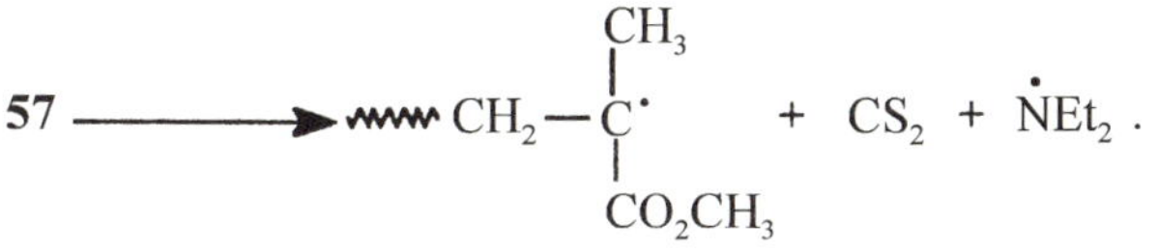

This side reaction could explain the loss of activity of the PMMA end-group. Besides, Van Kerckhoven et al. [237] have demonstrated that the quantum yield of the photodissociation of thiocarbamate-PMMA end-group is 2.5 times lower than that of thiocarbamate **51** and thiocarbamate-PS. Thus, all authors agree that side reactions occur with MMA-thiuram terminations, leading to a deactivation of the species.

Concerning acrylates, the radical polymerization of butyl acrylate has been extensively investigated by Lambrinos et al. [238], especially with XDT compound **60**. They have suggested that, among other side reactions, a disproportionation also occurs for the end-capped polyacrylates as follows:

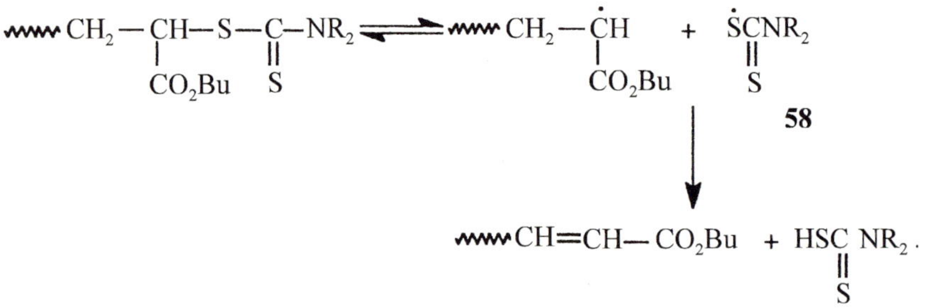

$$\sim\!\!\sim\!\!\sim CH_2-\underset{CO_2Bu}{CH}-S-\underset{S}{\overset{\|}{C}}-NR_2 \rightleftharpoons \sim\!\!\sim\!\!\sim CH_2-\underset{CO_2Bu}{\overset{\cdot}{CH}} + \underset{S}{\overset{\overset{\cdot}{\|}}{S}}CNR_2$$

58

$$\downarrow$$

$$\sim\!\!\sim\!\!\sim CH=CH-CO_2Bu + HS\underset{S}{\overset{\|}{C}}NR_2 .$$

59

This sulfurated product **59** leads to CS_2 and dialkylamine formation.

$$Et_2N\underset{S}{\overset{\|}{C}}SCH_2-\!\!\!\bigcirc\!\!\!-CH_2S\underset{S}{\overset{\|}{C}}NEt_2 \quad (XDT).$$

60

Their conclusion is based on the study of the distribution of SC(S) NEt$_2$ **58** groups on the polymer and on the dimer $Et_2NC(S)$ $SSC(S)$ NEt_2 as well as on the formation of by-products. The results show a decrease of the functionality from 2.0 to 0.5 once a conversion-rate of 75% is attained. At this stage, 37% of thiourea groups remain on the polymer whereas 30% of dimer were formed along with 33% of by-products. The authors have proposed several reactions to explain the loss of functionality: among them, disproportionation mentioned above, recombination of dithiocarbamyl **58** radicals, loss of CS_2 etc. However, none of these was completely satisfactory and the authors concluded that "the termination of the growing chains by dithiocarbonyl radicals may be considered as partly reversible".

The photochemical addition of butyl acrylate on polystyrene end-capped with **58** radical has been studied by Turner and Blevins [236]. They obtained a PS-*b*-PBuA diblock copolymer without formation of any homopolymer, in contrast to the case of styrene with MMA. This was confirmed by the inefficiency of ˙SCSNEt$_2$ radical **58** (produced by its dimer or the end-group) to initiate the homopolymerization. In the same way, Van Kerckhoven et al. [237] obtained from PS-S-C(S)NEt$_2$ as precursor, different diblock copolymers by reacting ethyl acrylate (EA), acrylic acid or MMA. In these cases, the living character was evidenced by an increase in molecular weight of the copolymers with the conversion. Similarly, the syntheses of PS-*b*-poly(acrylic acid) and PS-*b*-PMMA are also successful.

This behavior (reactivity of these acrylic comonomers), is in good agreement with that of classical radical copolymerization. However, the authors did not mention a loss of reactivity as observed in other studies of MMA.

3.2.4.5 Conclusion

Among the proposed compounds, thiurams lead to the most interesting results for the synthesis of block copolymers. Actually, the conclusion of all these investigations has been suggested by Doi et al. [239] who has recently summarized such research. After a survey of the photochemical polymerization of methyl acrylate (which is quite different from living polymerization), he has amalgamated the results described in the literature. The possible explanations of loss of the living character are summarized as

i) photodecomposition of $(M)_n SC(S)NR_2$ group along the polymerization,
ii) various cleavages of $\equiv$C–S–C(S)NR$_2$ end group, and
iii) bimolecular termination between macroradicals.

Doi et al. have convincingly demonstrated in the case of methyl acrylate that the two first hypotheses are not the right ones. On the other hand, the third is certainly the most likely as also suggested by Lambrinos et al. [238].

To circumvent this problem, Otsu has studied a system composed of the components dithiocarbamate **51** and thiuram disulfide **45** (where R = C$_2$H$_5$). The first produces the initiating radicals whereas the second is able to give a high amount of radicals according to the equilibrium

$$\left[\begin{array}{c} Et_2NCS \\ \| \\ S \end{array} \right]_2 \quad \rightleftharpoons \quad 2 \; \overset{\cdot}{S}CNEt_2 \atop \underset{S}{\|}.$$

Under these conditions, the radical polymerization appeared to be quasi living (i.e., Mn increases with monomer conversion) in contrast to reactions in the absence of thiuram disulfide. However, even if the authors did not mention it, it would be considered a drawback that the Mn value decreased because of the excess of terminating radical **58**. The proposed mechanism, where R represents macrogroups, is as follows:

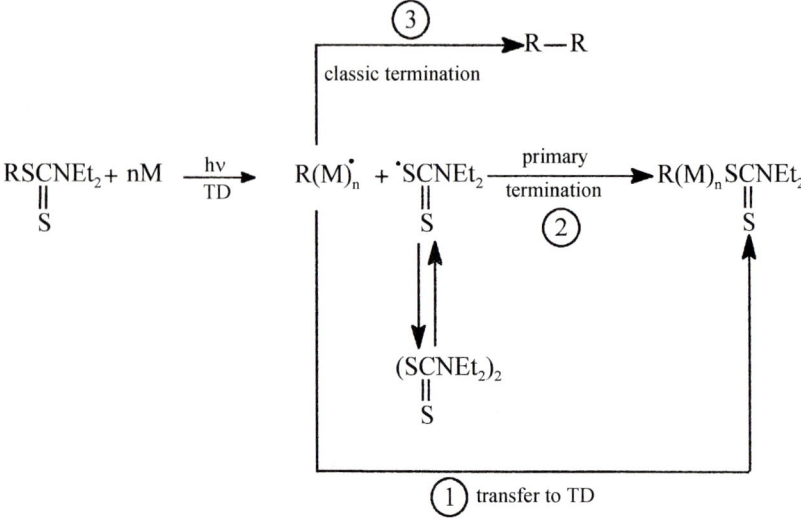

The advantage of such a new system allows one to reduce step ③ by an increase in the amount of active reactant thiuram disulfide which favors steps ① and ②. Here, the disadvantage of radical reactions can be found again, and similar behavior of counter radicals can also be observed for reactions involving nitroxyl radicals as recently described by Matyjaszewski et al. [62], Georges et al. [52] and Solomon [51] showing the evidence for the termination step.

Actually, at this stage, it is interesting to compare both thiuram and nitroxyl systems. The former react under UV irradiation preferentially whereas the latter gave satisfactory results with thermal initiation. Even if the dithiocarbamyl **58** counter radical is not so reactive, it is able to favor the polymerization of monomers whereas nitroxyls are totally unreactive since they are not able to initiate the polymerizations. These are the only two notable differences between both methods. However, according to the literature, it seems that thiuram disulfides can be used for acrylates and methacrylates more easily than in the case of nitroxyls.

3.2.5 Si–Si Cleavage

To our knowledge, only one study describes the synthesis of block copolymers from the cleavage of the Si–Si bond of silanes [240]. Poly(methyl phenyl silane) s **61** appear as efficient photoinitiators for the free radical polymerization of MMA leading, after short irradiation time, to PMMA **63** which reaches molecular weights of 54 000. These silylated PMMAs produce block copolymers **64** in fair yields up to 42% as follows:

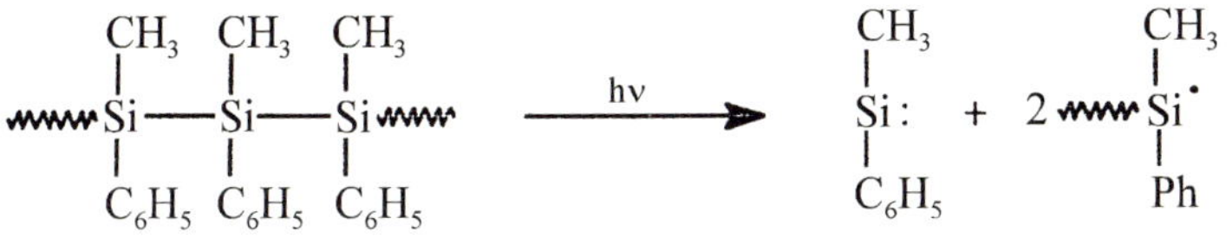

$$\text{wwwSi(CH}_3\text{)(C}_6\text{H}_5\text{)}-\text{Si(CH}_3\text{)(C}_6\text{H}_5\text{)}-\text{Si(CH}_3\text{)(C}_6\text{H}_5\text{)wwww} \xrightarrow{h\nu} \text{Si(CH}_3\text{)(C}_6\text{H}_5\text{):} + 2\text{ wwwSi(CH}_3\text{)(Ph)}^{\bullet}$$

61   **62**

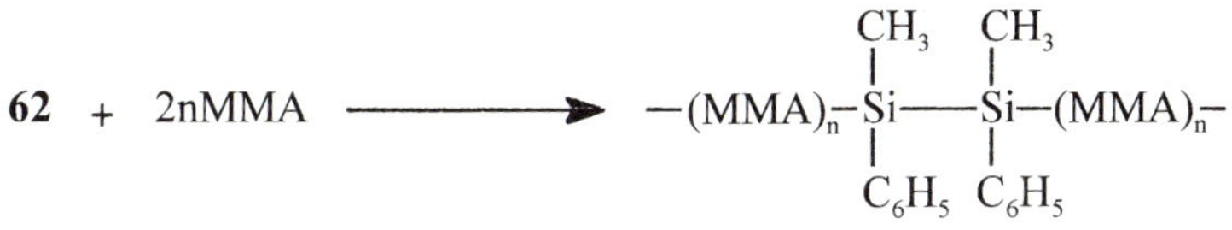

$$\mathbf{62} + 2n\text{MMA} \longrightarrow -(\text{MMA})_n-\text{Si(CH}_3\text{)(C}_6\text{H}_5\text{)}-\text{Si(CH}_3\text{)(C}_6\text{H}_5\text{)}-(\text{MMA})_n-$$

63

$$\mathbf{63} \xrightarrow{h\nu} 2-(\text{MMA})_n-\text{Si(CH}_3\text{)(C}_6\text{H}_5\text{)}^{\bullet} \xrightarrow{\text{Sty}} \text{PMMA}-b-\text{PS}-b-\text{PMMA}.$$

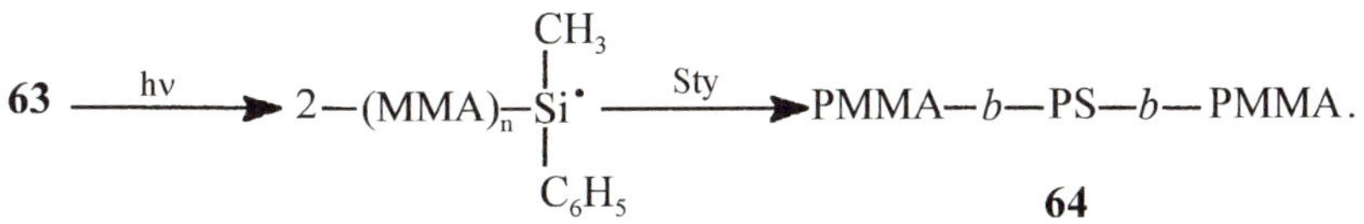

64

3.2.6 Conclusion

If an overall conclusion could be made, it might be considered that the counter-radicals vary considerably (Scheme 3). They can either be stable (e.g., nitroxyls, arylazooxyls), semi persistent (e.g., from thiourams) and also metallic (e.g., acetoacetato metals). In addition, if these radicals either terminate or transfer, non-living (or inactive) species will be produced. But, in order to preserve the living character, the radicals must propagate and in specific cases (e.g., iodine transfer polymerization or degenerative transfer) active species will be obtained. The more that one of these latter steps is favored, the more living is the tendency of the radical polymerization, with a very high kinetic control of this reaction.

4 Conclusion

This review has shown that several ways are possible to prepare block copolymers or cotelomers or cooligomers. However, according to the literature, the radical way is presently of increasing interest thanks to new processes, e.g., those involving counter radicals. Indeed, such a method does not require specific conditions such as high purity of monomer, inert atmosphere and strong

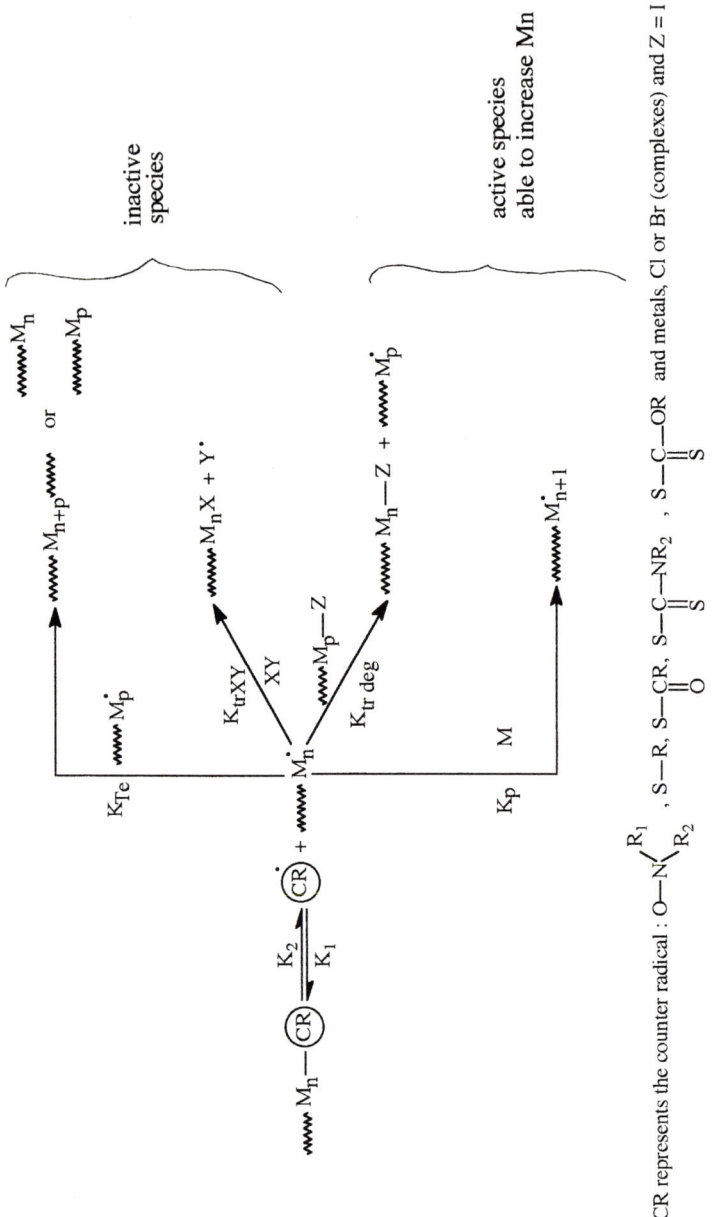

Scheme 3. Overall situation of radical controlled polymerization

vacuum. Even if radical polymerization initiated by macroradicals is still being investigated, more and more investigations involving the use of stable or persistent free radicals have focused the attention of many researchers onto the subject. The living character of the polymerization is of great interest.

Telomerization also appears as a powerful tool and opens up a wide range of opportunities to design tailored cooligomers and cotelomers. This rather soft process allows a high level of macromolecular engineering to be reached via radical conditions and the use of well-chosen and adequate telogens. Furthermore non-living and pseudo living behavior in telomerization is also possible. The first can be observed from the coupling of monofunctional oligomers or telomers or from the bistelomerization, whereas the living aspect of the telomerization has been shown by iodine transfer polymerization (in which the iodine atom acts similar to a nitroxyl group and this has already led to commercially available products with hard and soft segments) or degenerative transfer, or by the use of specific telogens (e.g., substituted tetraphenyl ethanes, telogens which exhibit S–S bonds or trityl group) that use the reversability of the termination step. Such new and sometimes recent developments show, to some extent, a rather good control of the steps of propagation and termination, and even if the initiation required a better control, these living radical polymerization could demonstrate a tendency to get closer to classical living ionic polymerizations.

However, even if radical telomerization is still a promising method, several problems still exist, requiring investigation especially of the reactivity control of the monomers and their radicals, the evidence of mechanism, the control of radicals, and also of the nature and the efficiency of the counter radicals, and the generalization to other monomers, particularly to dienes and halogenated olefins. In addition, knowledge of the molecular weights (especially in order to control the production of high molecular weight polymers) and of the molecular weight distributions, and sometimes the decrease of yields, mainly linked to the slow down of reaction rate owing to numerous steps of transfer-initiation have still to be extended and clarified. Furthermore, in the synthesis of block copolymers, the second and subsequent blocks also require a better control.

Such new targets are real challenges and should attract the interest of many academic and industrial researchers.

5 References

1. Noshay A, Mc Grath JE (1977). In: Block copolymers: overview and critical survey. Academic Press, New-York
2. Riess G, Hurtrez C, Bahadur P (1985) Block copolymers. In: Mark HF, Bikales NM, Overberger CG, Menges G (eds) Encyclopedia of polymer science and engineering, 2nd edn. Wiley, New York
3. Hazer B (1989) Synthesis and characterization of block copolymers. In: Cheremsinoff NP (ed) Handbook of polymer sciences and technology. Marcel Dekker, New York, Chap IV, p 133

4. Cowie JMG (1989) Block and graft copolymers. In: Allen G, Bevington JC, Eastmond AL, Russo A (eds) Comprehensive polymer science. Pergamon, Oxford, vol 3(3), p 33
5. Quirk RP, Kim J (1994) Block copolymers. In: Bloor D, Brook RJ, Flemings MC, Mahajan S (eds) Encyclopedia of advanced materials. Pergamon, Oxford, vol 1, p 280
6. Richards DH (1977) Chem Soc Rev 6: 235
7. Paul DR, Newman S (1978) Polymer Blends Vol 2, Academic Press, New York
8. Olabisi O, Robeson LM (1978) Polymer-Polymer Miscibility, Academic Press, New York
9. Lunsted LG, Schmolka IR (1976) In: Caresa RJ (ed), Block and Graft Copolymers, John Wiley & Sons Inc, New York, Vol 2
10. Yilgör J, Mc Grath JE (1988) Adv Polym Sc 86: 1
11. Li T, Topolkaraev VA, Holtner A, Baer E, Ji XZ, Quirk RP (1995) J Polym Sc Part B Polym Phys 34(4): 667
12. Reghunadhan Nair CP, Chaumont P, Clouet G (1994) Application of thermal iniferters in free radical polymerization: a new trend in macromolecular design In: Mishra MK (ed), Macromolecular Design: Concepts and Practice, Polymer Frontiers International Inc, Hopewell Jct, 11: 433
13. Simionescu CI, Comanita E, Pastravanu M, Dumitriu S (1986) Prog Polym Sci 12: 1
14. Jérome R, Henrioulle-Granville M, Boutevin B, Robin JJ (1991) Prog Polym Sci 16: 857
15. Ivanchev SS (1979) Polym Sci USSR 20: 2157
16. Boutevin B (1995) Contrôle de l'architecture des polymères par les radicaux libres In: Observatoire Français des Technologies Avancées (ed) Matériaux Polymères: Enjeux et Perspectives, Arago 16, Masson, Paris p 43
17. Moad G, Solomon DH (1989) Azo and Peroxy Initiators In: Allen G, Bevington JC, Eastmond AL, Russo A (eds), Comprehensive Polymer Science, Pergamon Press, Oxford 3(8): 97
18. Sheppard CS, Mc Leay RE, Bafford RA US Patent 4,088,642 (09-05-1978)
19. Sheppard CS, Mc Leay RE US Patent 4,101,522 (18-07-1978)
20. Schulz M, West G, Ourk S, Strunz J (1980) J Prakt Chem 322: 295
21. Boutevin B (1990) Adv Polym Sci 94: 70
22. Piirma I, Chou LPH (1979) J Appl Polym Sci 24: 2051
23. Gruber HF (1992) Prog Polym Sci 17: 953
24. Zhou G, Huang P, Zhu H (1995) Polym Bull 34: 295
25. Campbell DS, Mente PG, Tinkler AJ (1982) Rev Gen Caoutch Plast 619: 73
26. Campbell DS, Mente PG, Tinkler AJ (1983) Eur Patent 65,366 (24-11-1982) Chem Abst 98: 144415 k
27. Inoue H, Matsumoto A (1990) J Appl Polym Sci 40: 1917
28. Walz R, Bömer H, Heitz W (1977) Makromol Chem 178: 2527
29. Prisyazhnyuk SA, Ivanchev SS (1970) Polym Sci USSR 12 (2) : 514
30. Weber J, Thümmler W, Schnabel S, Schulz M (1982) German Patent 149,229 (01-07-1981) Chem Abst 96: 52961 e
31. Jimura K, Koide NO (1982) Rhe Proc IUPAC Makromol Symp 28th 825
32. Nuyken O, Gerum J, Kerber R (1980) Angew Makromol Chem 91: 143
33. D'Angelo AJ, Mageli OL (1976) US Patent 3,952,041 (20-04-1976), Chem Abst 85: 33928s
34. Simionescu CI, Popa AA, Comanita E, Pastravanu M (1990) Br Polym J 23: 347
35. Hazer B (1995) Macromol Chem Phys 196: 1945
36. Galibei VI, Arkhipova-Kalenchenko EG (1977) Zh Org Khim 13: 227
37. Nuyken O, Voit B (1994) Polymeric azo initiators In: Mishra MK (ed), Macromolecular Design: Concept and Practice, Polymer Frontiers International, Hopewell Jct (USA) Chapter 8: 13
38. Hazer B, Kurt A (1995) Eur Polym J 31: 499
39. Abadie MJM, Ourahmane D (1987) British Polym J 19: 247
40. Nicolova-Mankova Z, Palacin F, Raviola F, Riess G (1975) Eur Polym J 11: 301
41. Ger Patent 2,009,066 (1971)
42. Vinchon Y, Reeb R, Riess G (1976) Eur Polym J 12: 317
43. Reeb R, Vinchon Y, Riess G, Catala JM, Brossas J (1975) Bull Soc Chim France 11 12: 2717
44. Catala JM, Riess G, Brossas J (1977) Makromol Chem 178: 1249
45. Abadie MJM, Burgess FJ, Cunliffe AV, Richards DH (1976) J Polym Sci Poly Letters 14: 447
46. Richards DH (1985) ACS Symp Ser 87
47. Akar A, Aydogan AC, Talinli N, Yagci I (1986) Polym Bull 15: 293
48. Szwarc M, Levy M, Milkovich M (1956) J Am Chem Soc 78: 2656

49. Otsu T, Yoshida M (1982) Makromol Chem Rapid Commun 3: 127
50. Otsu T, Yoshida M, Tazaki T (1982) Makromol Chem Rapid Commun 3: 133
51. Solomon DH, Rizzardo E, Cacioli P (1984) Eur Pat Appl 0, 135, 280
52. Georges MK, Veregin RPN, Kazmaier PM, Hamer GK (1993) Macromolecules 26: 2987
53. Neri C, Costanzi S, Riva RM, Angaroni M (1991) Eur Pat Appl 0,488,403 A2
54. Ohkatsu Y, Yamaguchi K (1993) Polym Mat Sci Eng 68: 289
55. Nieman MB, Rozantzev EG, Mamedova Yu G (1962) Nature 100: 472
56. Vegerin RPN, Georges MK, Kazmaier PM, Hamer GK (1993) Macromolecules 26: 5316
57. Druliner JD (1991) Macromolecules 24: 6079
58. Druliner JD (1992) Adv Chem Ser 230: 95
59. Vegerin RPN, Georges MK, Hamer GK, Kazmaier PM (1995) Macromolecules 28: 4391
60. Georges MK, Veregin RPN, Kazmaier PM, Hamer GK (1993) Trends Polym Sci 2: 66
61. Georges MK, Veregin RPN, Hamer GK, Kazmaier PM (1994) Macromol Symp 88: 89
62. Matyjaszewski K, Gaynor S, Greszta D, Mardare D, Shigemoto T (1995) Macromol Symp 98: 73
63. Moad G, Rizzardo E, Solomon DH (1982) Macromolecules 15: 909
64. Grant RD, Rizzardo E, Solomon DH (1984) Makromol Chem 185: 1809
65. Rizzardo E, Serelis AK, Soloman DH (1982) Aust J Chem 35: 2013
66. Grant RD, Rizzardo E, Solomon DH (1985) J Chem Perkin Trans II 379
67. Grant RD, Griffiths PG, Moad G, Rizzardo E, Solomon DH (1983) Aust J Chem 36: 2447
68. Solomon DH, Rizzardo E, Cacioli P (1986) US Patent 4,581,429
69. Rizzardo E (1987) Chem Aust 54: 32
70. Keoshkerian B, Georges MK, Boils-Boissier D (1995) Macromolecules 28: 6381
71. Bertin D, Boutevin B (1996) Polym Bull under press; Bertin D, Boutevin B, Nicol P (1996) French Patent 96.05909 (To Elf Atochem SA) (13-05-1996)
72. Boutevin B, Cerf M, Pradel JL (1996) French Patent 96.06875 (to Elf Atochem SA) 04-06-1996
73. Mardare D, Matyjaszewski K (1993) Polym Preprint 34: 566
74. Mardare D, Matyjaszewski K (1994) Macromolecules 27: 645
75. Yamada B, Tanaka H, Konishi K, Otsu T (1994) Macromol Sci Chem A31: 351
76. Cadogan JIG, Paton RM (1971) J Chem Soc (B) 563
77. Kornblum N, Cheng L (1987) J Org Chem 52: 196
78. Hawker CJ, Carter KR, Hedrick JL, Voeksen W (1995) Polym Prepr 36(2): 110
79. Teyssié Ph (1994) Macromol Chem Macromol Symp 88: 1
80. Wayland BB, Poszmik G, Mukerjee SL, Fryd M (1994) J Amer Chem Soc 116: 7943
81. Mun Y, Sato T, Otsu T (1984) Makromol Chem 185: 1507
82. Mun Y, Sato T, Otsu T (1984) Makromol Chem 185: 1493
83. Gaynor S, Grezta D, Mardare D, Teodorescu M, Matyjaszewski K (1994) JMS Pure Appl Chem A31: 1561
84. Arvanitopoulos LD, Greuel MP, Harwood HJ (1994) ACS Polym Prepr 32(2): 549
85. Wayland BB, Fryd M, Mukerjee S, Posznik G (1995) Polym Mat Sc Eng 73: 422
86. Lee M, Minoura Y (1978) J Chem Soc Faraday Trans 1 74: 1726
87. Lee M, Ishida Y, Minoura Y (1982) J Poly Sc Poly Chem Ed 20: 457
88. Lee M, Morigami T, Minoura Y (1978) J Chem Soc Faraday Trans 1 74: 1738
89. Lee M, Utsumi K, Minoura Y (1979) J Chem Soc Faraday Trans 1 75: 1821
90. Hungenberg KD, Bandermann F (1983) Makromol Chem 184: 1423
91. Otsu T, Tazaki T, Yoshioka M (1990) Chemistry Express 5: 801
92. Yasuda H, Yamamoto H, Takemoto Y, Yamashita M, Yokota K, Miyake S, Nakamura A (1993) Makromol Chem Macromol Symp 67: 187
93. White D, Matyjaszewski K (1995) Polym Prepr 36: 286
94. Chung TC (1991) J Inorg Organometal Polym 1: 37
95. Chung TC (1988) Macromolecules 21: 865
96. Chung TC, Jiang GJ (1992) Macromolecules 25: 4816
97. Chung TC, Janvikul W, Bernard R, Jiang GJ (1994) Macromolecules 27: 26
98. Jiang GJ, Cheng JY, Lin CH (1995) Polym Mater Sc Eng (1995) 73: 542
99. Bamford CH (1974) Reactivity, Mechanism and Structure of Polymer Chemistry, Jenkins AD, Ledwith A (Ed) New York, chap 3
100. Bamford CH, Han X (1981) Polymer Comm 22: 1299
101. Niwa M, Katsurada N, Higashi N (1988) Macromolecules 21: 1878
102. Kang YJ, Soo SG, Kwon RJ (1980) Pollino 4(5): 392 (Chem Abst 94: 66145)
103. Asscher M, Vofsi D (1968) J Chem Soc 947

104. Freidlina R Kh, Chukovskaya E Ts (1973) Izv Akad Nauk SSSR 8: 1782
105. Boutevin B, Maubert C, Pietrasanta Y, Sierra P (1981) J Polym Sci Part A Polym Chem 19: 511
106. Boutevin B, Maubert C, Pietrasanta Y, Mebkhout A (1981) J Polym Sci Part A Polym Chem 19: 499
107. Battais A, Boutevin B, Hugon JP, Pietrasanta Y (1980) J Fluorine Chem 16: 397
108. Améduri B, Boutevin B, Lecrom C, Garnier L (1992) J Polym Sci Part A Poly Chem 30: 49
109. Boutevin B, Pietrasanta Y, Taha M (1982) Makromol Chem 184: 2401
110. Boutevin B, Maliszewicz M, Pietrasanta Y (1983) Makromol Chem 184: 977
111. Boutevin B, Macret M, Maubert C, Pietrasanta Y, Tanesie M (1978) Tetrahedron Letters 33: 3019
112. Park JH, Sur SG, Sam K (1980) Taehan Hwahakkoe Chi 24(3): 259 (Chem Abst 94: 4265 d)
113. Sherman PO, Smith S (1968) French Patent 1, 562, 070 (to Minnesota Mining and Manufacturing Co) (26.01.1968)
114. Erickson JG (1967) US Patent 3,278,352 (to 3M) (Chem Abst 66: 56380)
115. Boutevin B, Pietrasanta Y (1988) In: Erec (Ed) Les Acrylates et Polyacrylates Fluorés: Dérivés et Applications, Paris
116. Dessaint A, Perronin J (1977) French patent 2,328,070 and (1980) French patent 2,442,861 (to Produits Chimiques Ugine Kulhman)
117. Asahi Glass (1981) Japanese Patent 81,49,348 of 02.05.81 (Chem Abst 95: 97103h)
118. Tung LH, Lo GYS, Griggs JA (1985) J Polym Sci Part A Polym Chem Ed 23: 1551
119. Eastmond GC, Parr KJ, Woo J (1988) Polymer 29: 950
120. Sato T, Terada K, Yamauchi J, Okaya T (1993) Makromol Chem 194: 175
121. Uraneck CA (1969) J Appl Polym Sci 13: 149
122. Rempp P, Franta E, Herz JE (1988) Adv Poly Sc 86: 145
123. Haesslin HW (1985) Makromol Chem 186: 357
124. Chaumont P, Beinert G, Herz JE, Rempp P (1981) Polymer 22: 663
125. Gagnebien D, Madec PJ, Marechal E (1985) Eur Poly J 21: 301
126. Bachari A, Belorgey G, Hélary G, Sauvet G (1995) Macromol Chem Phys 196: 411
127. Boutevin B, Lusinchi JM, Pietrasanta Y, Robin JJ (1994) Eur Polym J 30(5) : 615
128. Esselborn E, Fock J, Knebelkamp A (1996) Macromol Symp 102: 91
129. Kennedy JP, Hongu Y (1985) Polym Bull 13: 115
130. Japanese Patent 196,583 of 5.03.94 (Chem Abst 104: 178299s)
131. Okano T, Katayama M, Shinohara I (1978) J Applied Polym Sci 22: 369
132. Shimada M, Miyahara M, Tahara H, Shinohara I, Okano T, Kataoka K, Sakurai Y (1983) Polymer J 15(9): 649
133. Tezuka Y, Imai K (1984) Makromol Chem Rapid Comm 5: 559
134. Boutevin B, Fleury E, Pietrasanta Y, Sarraf L (1986) Polym Bull 15: 107
135. Wang JS, Matyjaszewski K (1995) Polym Mat Science Eng 73: 414
136. Percec V, Barboin B (1995) Macromolecules 28: 7970
137. Kato M, Kamigaito M, Sawamoto M, Higashimura T (1995) Macromolecules 28: 1721
138. Wang JS, Matyjaszewski K (1995) J Amer Chem Soc 117: 5614
139. Wang JS, Greszta D, Matyjaszewski K (1995) Polym Mat Science Eng 73: 416
140. Sawamoto M, Kato M, Kamigaito M, Higashimura T (1995) Poly Preprints 36: 539
141. Matsumoto H, Nakano T, Nagai Y (1973) Tetrahedron Lett 51: 5147
142. Matsumoto H, Nikaido T, Nagai Y (1976) J Org Chem 41: 396
143. Sasson Y, Rempel GL (1975) Synthesis 448
144. Améduri B, Boutevin B (1990) Macromolecules 23: 2433
145. Améduri B, Boutevin B (1991) Macromolecules 24: 2475
146. Kameda N, Ishii E (1983) Makromol Chem 184: 1901
147. Kameda N, Ishii E (1983) Nippon Kagaku Kaishu 8: 1196 (Chem Abst 99: 123010)
148. Améduri B, Boutevin B, unpublished results
149. Kotora M, Kvicala J, Améduri B, Hajek M, Boutevin B (1993) J Fluorine Chem 64: 259
150. Tatemoto M (1990) Kagaku Kogyo 41(1): 78 (Chem Abst 114: 8081v)
151. Tatemoto M (1992) Kobunshi Ronbunshu 49(10): 765 (Chem Abst 118: 22655z)
152. Tatemoto M, Tomoda M, Ueta Y (1980) Ger Patent DE 29,401,35 (to Daikin Kogyo Co Ltd Jap) (Chem Abst 93: 27580)
153. Tatemoto M, Morita S (1981) Eur Pat Appl EP 27,721 (to Daikin Kogyo Co Ltd) (Chem Abst 95: 170754)
154. Tatemoto M, Suzuki T, Tomoda M, Furukawa Y, Ueta Y (1978) Ger Patent DE 2,815,187 (to Daikin Kogyo Co Ltd Jap) (Chem Abst 90: 24564)

155. Tatemoto M (1978) Jap Patent 53,026,781 (to Daikin Kogyo Co Ltd Jap) (Chem Abst 89: 111170)
156. Tatemoto M, Furukawa Y, Tomoda M, Oka M, Morita S (1980) Eur Pat Appl EP 14,930 (to Daikin Kogyo Co Ltd Jap) (Chem Abst 94: 48603)
157. Hanford WE, Joyce Jr RM (1949) US Patent 2,440,800
158. Haszeldine RN (1949) J Chem Soc 2859
159. Starks CM (1974) Free Radical Telomerization 1st Ed, Academic Press, New-York
160. Boutevin B, Pietrasanta Y (1989) Telomerization In: Allen G, Bevington JC, Eastmond AL, Russo A (Eds) Comprehensive Polymer Sc, Pergamon, Oxford 3: 185
161. Gordon R, Loftus RD (1989) Telomerization In: Kirk RE, Othmer DF (Eds) Ency Poly Sci Tech, Wiley 16: 533
162. Améduri B, Boutevin B (1994) Telomerization In: Bloor D, Brook RJ, Flemings RD, Mahajan S (Eds) Encyclopedia Adv Materials Pergamon, Oxford 2767
163. Tatemoto M, Nakagawa T (1975) Jap Patent 50,127,991 (to Daikin Kogyo Co Ltd Jap) (Chem Abst 84: 136949)
164. Daikin Ind Ltd (1994) Fluoroelastomer Catalogue
165. Tatemoto M (1985) Int Poly Sc Tec 12(4): 85 translated in english from (1984) Nippon Gomu Kyokaishi 57(11): 761
166. Oka M, Tatemoto M (1984) Contemp Topics Polym Sci 4: 763
167. Tatemoto M (1979) "Recent studies on fluoro-elastomers of vinylidene fluoride". First regular meeting of soviet-japanese fluorine chemists, Tokyo 15–16 Feb.
168. Tatemoto M, Nakagawa T (1977) Ger Patent DE 2,729,671 (to Daikin Kogyo Co Ltd Jap) (Chem Abst 88: 137374)
169. Tatemoto M (1994) Porima Daijesuto 46(6): 61 (Chem Abst 121: 85440)
170. Tatemoto M (1990) Eur Patent Appl EP 399,543 (to Daikin Ind Ltd Jap) (Chem Abst 114: 166150)
171. Tatemoto M, Yagi T (1988) Eur Patent Appl EP 268,157 (to Daikin Ind Ltd Jap) (Chem Abst 110: 13632)
172. Tatemoto M, Nakagawa T (1987) Jap Patent 87,021,805 (to Daikin Ind Ltd Jap) (Chem Abst 107: 177767)
173. Vergé C (1991) Thèse de l'Université de Montpellier II, France
174. Bauduin G, Boutevin B, Lantz A, Vergé C, J Fluorine Chem (in press)
175. Chambers RD, Greenhall MP, Wright AP, Caporiccio G (1995) J Fluorine Chem 73: 87
176. Gornowicz GA, Boutevein B, Caporiccio G (1993) US Patent 5,196,614 (to Dow Corning Corp) 23-04-93
177. Améduri B, Boutevin B, Kostov GK, Petrova P (1995) J Fluorine Chem 74: 261
178. Boutevin B, Gornowicz GA, Caporiccio G (1992) US Patent 3,671 (to Dow Corning Corp) 05-08-92
179. Balagué J, Améduri B, Boutevin B, Caporiccio G, J Fluorine Chem (in press)
180. Balagué J (1994) Thèse de l'Université de Montpellier II, France
181. Balagué J, Améduri B, Boutevin B, Caporiccio G submitted to J Fluorine Chem
182. Balagué J, Améduri B, Boutevin B, Caporiccio G submitted to J Fluorine Chem
183. Balagué J, Améduri B, Boutevin B, Caporiccio G (1995) J Fluorine Chem 70: 215
184. Balagué J, Améduri B, Boutevin B, Caporiccio G (1995) J Fluorine Chem 73: 237
185. Balagué J, Améduri B, Boutevin B, Caporiccio G (1995) J Fluorine Chem 74: 49
186. Apsey GS, Chambers RD, Salisbury MJ, Moggi G (1988) J Fluorine Chem 40: 261
187. Tortelli V, Tonelli C (1990) J Fluorine Chem 47: 199
188. Baum K, Archibald TG, Malik AA (1993) US Patent 5,204,441 (to Fluorochem Inc)
189. Baum K, Malik A (1994) J Org Chem 59: 6804
190. Manséri A, Améduri B, Boutevin B, Chambers RD, Caporiccio G, Wright AP (1995) J Fluorine Chem 74: 59
191. Manséri A, Améduri B, Boutevin B, Chambers RD, Caporiccio G, Wright AP submitted to J Fluorine Chem
192. Manséri A (1994) Thése de l'Université de Montpellier II, France
193. Manséri A, Améduri B, Boutevin B, Caporiccio G, J Fluorine Chem (in press)
194. Boutevin B, Caporiccio G, Guida-Pietrasanta F, Ratsimihéty A (1995) Italian Patent M 95/A 000701 (to Dow Corning Corp) 06-04-95
195. Matyjaszewski K, Gaynor SG, Greszta D, Mardare D, Shigemoto T, Wang JS (1995) Macromol Symp 95: 217
196. Gaynor SC, Wang JS, Matyjaszewski K (1995) Polym Preprints 36: 467

197. Wang JS, Gaynor SC, Matyjaszewski K (1995) Polym Preprints 36: 465
198. Borsig E, Lazar M, Capla M (1967) Makromol Chem 105: 212
199. Borsig E, Lazar M, Capla M (1967) Collect Czech Chem Commun 32: 4289
200. Braun D (1986) Makromol Chem Macromol Symp 4: 41
201. Bledzki A, Braun D, Titzschkau K (1987) Makromol Chem 188: 2061
202. Bledzki A (1983) Kunstoffe 73(3): 156
203. Bledzki A, Braun D, Menzel W, Titzschkau K (1983) Makromol Chem 184: 287 and 745
204. Braun D, Skrzek T, Steinhauer-Beisser S, Tretner H, Lindner HJ (1995) Macromol Chem Phys 196: 573
205. Braun D (1994) Angew Makromol Chem 223: 69
206. Bledzki A, Braun D (1986) Polym Bull 16: 19
207. Bledzki A, Braun D (1986) Makromol Chem 187: 2599
208. Braun D, Skrzet T (1995) Makromol Chem Phys 196: 4039
209. Auwers K, Meyers V (1889) Ber 22: 1228
210. Constanza AJ, Coleman RJ, Pierson RM, Marvel CS, King C (1955) J Polym Sci 24: 319
211. Crivello JV, Lee JL, Conlon DA (1986) J Polym Sci Part A Polym Chem 24: 1251
212. Crivello JV, Conlon DA, Lee JL (1986) J Polym Sci Part A Polym Chem 24: 1197
213. Otsu T, Matsumoto A (1994) Macroiniferters: controlled synthesis and design through living radical polymerization" In: Mishra MK (Ed) Macromolecular Design: Concept and Practice, Polymer Frontier International, Hopewell Jct (USA) Chapter 12: 471
214. Otsu T, Takazi T (1986) Polym Bull 16: 277
215. Tortelli V, Tonelli C Int. Patents WO8,808,007 A1 881020 and WO8901927 A1 890309 (to Ausimont)
216. Tortelli V, Tonelli C, Corvaja C (1993) J Fluorine Chem 60: 165
217. Tonelli C, Tortelli V (1994) J Fluorine Chem 67: 125
218. Demircioglu P, Acar MH, Yagci Y (1992) J Appl Poly Sci 46: 1639
219. Yagci Y, Mishra MK (1994) Macroiniferters for chain polymerization In: Mishra MK (Ed) Macromolecular Design: Concept and Practice Polymer Frontiers International Hopewell Jct, Chapter 6: 229
220. Ferrington TE, Tobolsky AV (1955) J Amer Chem Soc 77: 4510
221. Kennedy JP (1979) J Macromol Science Chem 13: 695
222. Otsu T (1956) J Polym Science 21(99): 559
223. Otsu T, Nayatani K, Muto I, Ima M (1958) Makromol Chem 27: 142
224. Staudner E, Kyiela G, Beniska J, Mikolaj A (1978) Eur Polym J 14: 1067
225. Kumar RC, Dueltgen RR, Andrus MH (1994) Polymeric Photoiniferters In: Mishra MK (Ed) Macromolecular Design: Concept and Practice, Polymer Frontiers International, Hopewell Jct (USA) chapter 13: 487
226. Akemi H, Aoyagi T, Shinohara I, Okano T, Katoaka K, Sakurai Y (1986) Makromol Chem 187: 1627
227. Skok A, Opresnik M, Sebenik A, Osredkar U (1994) Macromolecular Reports A31 (Suppl 6 and 7): 1263
228. Opresnik M, Sebenik A (1995) Polym Inter 36: 13
229. Fokina TA, Apukhtina NP, Skii ALK, Nelson KN, Solodovnikova GS (1966) Vysokomol Soyed 8(12): 2197
230. Niwa M, Matsumoto T, Izumi H (1987) J Macromol Science Chem 24: 567
231. Niwa M, Sako Y, Shimizu M (1987) J Macromol Science Chem A 24 (11): 1315
232. Reghunadhan Nair CP, Richou MC, Clouet G, Chaumont P (1990) Eur Polym J 26: 811
233. Reghunadhan Nair CP, Richou MC, Clouet G, Chaumont P (1991) Makromol Chem 192: 579
234. Imoto M, Otsu T, Yonezawa J (1960) Macromol Chem 36: 93
235. Otsu T, Yoshida M, Kuriyama A (1982) Polym Bull 7: 45
236. Turner SR, Blevins RW (1990) Macromolecules 23: 1856
237. Van Kerckhoven C, Van der Brueck H, Smets G, Huybrechts J (1991) Makromol Chem 192: 101
238. Lambrinos P, Tardi M, Polton A, Sigwalt P (1990) Eur Polym J 26(10): 1125
239. Doi T, Matsumoto A, Otsu T (1994) J Polym Sci Part A Polym Chem 32: 2918
240. Yucesan D, Hostoygar H, Denizligil S, Yagci Y (1994) Angew Makromol Chem 221: 207

Editor: Prof. L. Monnerie
Received: March 1996

Metallocenes for Polymer Catalysis

Walter Kaminsky and Michael Arndt
Universität Hamburg, Institut für Technische und Makromolekulare Chemie
Bundesstr. 45, 20146 Hamburg Germany

List of Abbreviations		144
1	**Introduction**	145
2	**Function of the Cocatalyst**	147
3	**Ethene Polymerization**	149
	3.1 Homopolymers	149
	3.2 Copolymers	154
	3.3 Long Chain Branching	156
4	**Polypropene**	157
	4.1 Regio- and Stereospecificity	157
	4.1.1 Microstructure of Polypropene	157
	4.1.2 Symmetry and Stereospecificity	159
	4.2 Oligomerization and Mechanism	165
	4.3 Polymer Properties	167
	4.3.1 Atactic Polypropene	167
	4.3.2 Isotactic Polypropene	167
	4.3.3 Syndiotactic Polypropene	170
	4.3.4 Elastomeric Polypropene	170
	4.4 Heterogenization and Polymerization in the Presence of Fillers	171
	4.4.1 Synthesis of a Metallocene on a Carrier	172
	4.4.2 Attaching a Metallocene to a Carrier	172
	4.4.3 Fixing a Metallocene on Supported MAO	172
5	**Cycloolefin Polymerization**	174
	5.1 Homopolymers	174
	5.2 Copolymers (COC)	176
6	**Other Monomers**	178
	6.1 Styrene	178
	6.2 Dienes	179
	6.3 Methylmethacrylate	181
7	**Conclusion**	182
8	**References**	182

List of Abbreviations

*t*Bu	*tert*-Butyl
Bz	Benzyl
Cp	Cyclopentadienyl
En	1,2 Ethanediyl
Flu	9-Fluorenyl
Ind	1-Indenyl
IndH4	4,5,7,8-Tetrahydro-1-indenyl
Me	Methyl
Naph	Naphthyl
Ph	Phenyl
iPr	iso-Propyl
TMA	Trimethylaluminium
TEA	Triethylaluminium
TIBA	Triisobutylaluminium
MAO	Methylaluminoxane
EAO	Ethylaluminoxane
iBAO	Isobutylaluminoxane
aPP	atactic polypropene
iPP	isotactic polypropene
sPP	syndiotactic polypropene

1 Introduction

The polymerization of olefins to polymers with different microstructures and properties continues to be one of the most investigated areas for both industrial and academic laboratories in polymer science [1–5]. The usage of polyolefins, especially polypropylene, as polymer materials is rapidly growing and reaches a production level of 50 million tons a year. It is estimated for the year 2005 that polyolefins will form 55% of the whole production of plastics [6]. There are three reasons: (i) polyolefins are made from simple and easily available compounds (ethene, propene, diene), given by crude oil processes; (ii) they are known for their low energy demand during polymerization and melt processing; (iii) they contain only hydrogen and carbon and can be reused by granulation or degraded by thermal processes to oil and monomers, or used as an energy source for incineration.

These properties are the driving forces in the development of new polyolefins which aims at broadening the properties envelopment of polyolefins and expanding their boundaries towards the areas traditionally occupied by more sophisticated, expensive and sometimes hazardous materials. Forty years after the important discovery of the metallorganic catalyzed polymerization of olefins by Karl Ziegler [7] and the stereospecific polymerization of propene and α-olefins by Giulio Natta [8], the use of metallocene catalysts shows the way to expand the possibilities of olefin polymerization and the properties of the resulting polyolefin materials. This new generation of catalysts offers a versatile method in the synthesis of polymers, a better mechanistic understanding of the polymerization steps, and great electronic and steric variations in the cyclopentadienyl-type ligands for modelling in the design of catalytic systems. Scientific understandings in this area have reached a level where tailoring of polyolefins has become a reality [2, 9, 10].

Metallocene catalysts are bicomponents consisting of group four transition metal compounds and cocatalysts. Metallocene compounds have long been known and were used as a compound for Ziegler-Natta catalysts (Table 1) [11].

The traditional cocatalyst, diethylaluminumchloride or triethylaluminum, shows only a pure polymerization activity and was used as a homogeneous system to understand the polymerization, which is simpler in soluble than in heterogeneous systems. Kinetic studies and applications of various methods have helped to define the nature of the active centers, to explain aging effects, to establish the mechanism of the interaction with olefins, and to obtain quantitative evidence of some elementary steps [12, 13].

Breslow and Newburg [14] used bis-cyclopentadienyltitaniumdichloride (Cp_2TiCl_2) together with diethylaluminumchloride for the polymerization of ethene. Subsequent research on this and other metallocene systems with various alkyl groups has been performed by Natta et al. [15], Belov et al. [16], Dyachkovskii et al. [17], Patat and Sinn [18], Chien and Hsieh [19], Clauss and Blstian [20], Henrici-Olivé and Olivé [21], Reichert and Schoetter [22], and

Table 1. Timetable and historical development of metallocene research

1952	Development of the structure of metallocenes (ferrocene) Fischer, Wilkinson
1955	Metallocene as component of Ziegler-Natta catalysts, low activity with common aluminium alkyls Natta, Breslow
1973	Addition of small amount of water to increase the activity (Al:$H_2O \sim 1:0.05$ up to $1:0.3$) Reichert, Meyer, Long, Breslow
1975	Unusual increase of activity by adding water at the ratio Al:$H_2O = 1:2$ Kaminsky, Sinn, Mottweiler
1977	Using separate prepared methylaluminoxane (MAO) as cocatalyst together with metallocenes for olefin polymerization Kaminsky, Sinn
1982	Synthesis of ansa metallocenes with C2-symmetry Brintzinger
1984	Polymerization of propene using rac/meso mixture of ansa titanocenes lead to parts of isotactic polypropylene Ewen
1984	Chiral ansa zirconocenes produce highly isotactic polypropylene Kaminsky, Brintzinger

Fink [23]. It was shown that, after a short induction period, the activity of polymerization reaches a maximum, then decreases continuously, due to fast aging processes such as alkyl exchange, hydrogen transfer, and reduction of the transition metal species.

A remarkable increase in activity (factor 20–100) was found by Reichert and Meyer [24] by adding small amounts of water (alkyl:$H_2O = 1:20$) to the system $Cp_2TiCl_2/C_2H_5AlCl_2$, and by Long and Breslow [25] to the catalyst Cp_2TiCl_2/Me_2AlCl. These workers proposed a stabilized catalyst complex resulting from an increase in Lewis acidity. Before this, water was considered to be a catalyst poison. Only some patents have been granted for adding water for purposes of lowering the molecular weights and improving the molecular mass distribution.

An enormous increase (factor up to 1 million) in activity was found in 1975 at the University of Hamburg when water was added in a ratio of $Al(CH_3)_3$: $H_2O = 1:2$ and, in 1977, using the isolated reaction product of methylaluminoxane (MAO) together with titanocenes and zirconocenes ($Cp_2Ti(CH_3)_2$, $Cp_2Zr(CH_3)_2$, Cp_2ZrCl_2) as catalysts for ethene polymerization [26, 27]. In these combinations, metallocenes become more active than commercially used Ziegler catalysts.

The next important step was made using the ansa metallocenes synthesized by Brintzinger et al. in 1982 [28] for the stereospecific polymerization of propene. Ewen et al. [29] succeeded 1988 in synthesizing a Cs-symmetric zirconocene ($[Me_2C(Flu)(Cp)]ZrCl_2$) which produces syndiotactic polypropylene in high quantities. Since 1985, a rapid, worldwide industrial and academic development began in the field of metallocene catalysts which continues today.

2 Function of the Cocatalyst

A key to the high polymerization activity of metallocenes are the cocatalysts. Methylaluminoxane (MAO) is mostly used and is synthesized by controlled hydrolysis of trimethyl aluminium [30]. Other bulky anionic complexes which show a weak coordination, such as borates, play an increasing role too. One function of MAO is the alkylation of halogenated metallocene complexes. In the first step, the monomethyl compound is formed within seconds even at $-60\,°C$ [31]. Excess MAO leads to the dialkylated species as NMR measurements show. In order for the active site of form, it is atleast necessary that one alkyl group is bonded to the metallocene [32].

$$Cp_2M\begin{matrix}Cl\\Cl\end{matrix} \xrightarrow{MAO\ or\ TMA} Cp_2M\begin{matrix}Me\\Cl\end{matrix} \xrightarrow{MAO} Cp_2M\begin{matrix}Me\\Me\end{matrix}$$

The structure of MAO is complex and has been investigated by cryoscopic measurements, hydrolysis reactions, IR-, UV-, mass, and NMR-spectroscopic measurements and other methods [33–39]. There are different equilibria between the oligomers. The molecular weight determined cryoscopically in benzene lies between 1000 and 1500 g/mol. Since the work of Sinn [40] and Barron [41], more details about the structure of methylaluminoxane or tert-butylaluminoxane have become known. Amongst the different oligomers, $(CH_3)_2Al-O-Al(CH_3)-OAl(CH_3)_2$ units are important. These units can associate resulting in coordination of the unsaturated aluminum atoms. There are tri- and tetra-coordinated aluminium atoms, of which the trivalent show extreme Lewis acidity. Four of the $Al_4O_3(CH_3)_6$ units can form a cage structure resembling a half open dodecaeder ($Al_{16}O_{12}(CH_3)_{24}$). This cage is complexed with differing amounts of trimethylaluminium. The MAO complex can seize a methyl anion, a Cl^- anion or an OR^- anion from the metallocene forming an AlL_4- anion which can distribute the electron over the whole cage, thus stabilizing the charged system.

$$Cp_2M\begin{matrix}Me\\Me\end{matrix} \rightleftarrows Cp_2M\begin{matrix}Me\\Me\end{matrix}\cdots Al_{MAO} \rightleftarrows Cp_2M\begin{matrix}Me\\\oplus\end{matrix}\quad Me\ ^\ominus AlMAO$$

There is much evidence that the cationic $L_2M(CH_3)^+$ is the active center in olefin polymerization.

The appearance of cationic metallocene complexes $L_2M(CH_3)^+$, especially $Cp_2Zr\ CH_3^+$ in the presence of MAO-solution, was confirmed by ^{13}C-NMR, 91Zr-NMR and X-ray measurements [42–44]. The cation is stabilized by a weak coordination to Al–O units or bridging CH_3, Cl ligands. Further evidence for the existence of cationic centers is given by the activation of metallocene catalysts for olefin polymerization using anionic counterions such as tetraphenylborate $(C_6H_5)_4B^-$, carborane $(C_2B_9H_{12}^-)$ or fluorinated borate. The use of $(C_6F_5)_4B^-$ by Hlatky et al. [45], Marks et al. [46], and Zambelli et al. [47] as counter ion leads to highly active metallocene catalysts. They are formed by the reaction of a dialkylated zirconocene with dimethylaniliniumtetrakis(pentafluorophenyl) borate.

$$Cp_2ZrMe_2 + [NHMe_2Ph][B(C_6F_5)_4] \longrightarrow$$

$$[Cp_2ZrMe]^+ + [B(C_6F_5)_4]^- + NMe_2Ph + CH_4$$

Cationic metallocene complexes can also be formed by reactions of perfluorinated triphenylborane or trityltetrakis(pentafluorophenyl) borate.

$$Cp_2ZrMe_2 + B(C_6F_5)_3 \longrightarrow [Cp_2ZrMe]^+ + [MeB(C_6F_5)_3]^-$$

$$Cp_2ZrMe_2 + [Ph_3C][B(C_6F_5)_4] \longrightarrow$$

$$[Cp_2ZrMe]^+ + [B(C_6F_5)_4]^- + Ph_3CMe$$

X-ray analysis of solid compounds such as $[(CH_3)_2C_5H_3]ZrCH_3^+ \leftrightarrow H_3CB(C_6F_5)_3]^-$ by Yang et al. [48] shows that a part of a coordination bond still exists between the zirconocene and the borate. The olefin π-complex bonded by this compound and then inserted into the zirconium-methyl bond.

A great advantage is that the ratio of borate to metallocene is about 1 to 1 and not 5000 to 1, as in the case of MAO. On the other hand, the borate system is very sensitive to poisons and decomposition and must be stabilized by addition of aluminiumalkyls such as tri-isobutylaluminium [49–51].

Another function of MAO is the reactivation of inactive complexes formed by hydrogen transfer reactions. If metallocenes and MAO are combined in solution, methane and an catalytically inactive M–CH$_2$–Al complex are formed after fast complexation and methylation. MAO seems to be a strong Lewis acid which shows a high rate of α-hydrogen transfer (Fig. 1) [52].

The rate of methane production depends on the Al/Zr ratio, temperature and type of metallocene. Upon enlargement of the molar ratio of Al/Zr, the rate of methane evolution increases to reach a level of 1.0 mol methane per mol of zirconocene in 150 min at Al/Zr = 420. That means that every zirconocene must have reacted and formed an inactive complex. In spite of this, strong polymerization activity is still observed. The explanation for this is a reactivation of inactive structures by MAO via ligand exchange.

Zr–CH$_2$–Al-structures react with MAO and form Zr–CH$_3$ and Al–CH$_2$–Al structures. Even after 20 h reaction time, a small concentration of Zr–CH$_3$ can

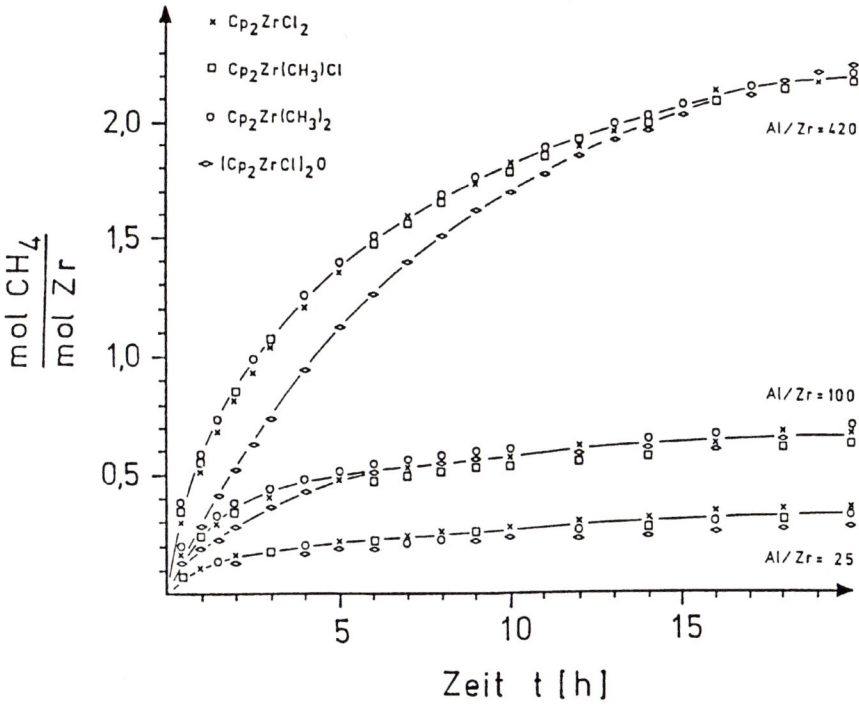

Fig. 1. Time dependence of rate of methane production in the reaction of various zirconocenes with methylaluminoxane at different Zr/Al ratios

be identified by ^{13}C-NMR measurements. This is one explanation why a high excess of MAO is necessary to reach an optimal polymerization activity [53].

3 Ethene Polymerization

3.1 Homopolymers

The zirconocene/MAO catalysts are about 10–100 times more active for ethene polymerization than the conventional Ziegler systems. Using bis(cyclopentadienyl) zirconium dichloride Cp_2ZrCl_2 and MAO up to 40 000 000 g, polyethene/g Zr·h are obtained (Table 2) [54].

Every zirconium atom forms an active complex as shown by Tait [55] and Chien and Wang [56] and produces about 46 000 polymer chains per hour. The time of insertion of one ethene unit is only 3×10^{-5} seconds. There are similar rates as they were observed for enzymes. The analogy can be seen in many other fields too (influence of substitution, regioselectivity, stereospecificity). As

Table 2. Polymerization activity of [Cp$_2$ZrCl$_2$]/MAO catalyst applied to ethene in 330 ml of toluene

[Cp$_2$ZrCl$_2$]	6.2×10^{-8} mol/l
[MAO]	7.1×10^{-4} mol/l
Activity (95 °C, 8 bar)	39.8×10^3 kgPE/(molZr × h)
M$_n$ of polyethene	78 000 g/mol
P$_n$ of polyethene	2800
Macromolecules per Zr-atom and hour	46 000
Rate of growth of one macromolecule	0.087 s
Turnover time	3.1×10^{-5} s

metallocenes for ethene polymerization, unbridged, bridged, substituted and half-sandwich complexes have been used.

Figure 2 shows some of the classes of metallocene catalysts used for the polymerization of ethene. In order to compare the reactivities and molecular masses, the polymerizations are carried out under the same conditions (30 °C, 2 bar ethene pressure, toluene as solvent) or calculated to these parameters by data from the literature [57–60].

Table 3 shows the polymerization behaviour of different metallocene aluminoxane catalysts. Generally, zirconium catalysts are more active than the hafnium or titanium systems. Especially partially substituted bisindenyl systems **24** show very high activities, exceeding those of the sterically less hindered Cp$_2$ZrCl$_2$ **1**. Zirconocenes with bulky ligands such as neomenthyl substituted derivative **12** afford significantly lower productivity. This indicates that electron-donating groups can enhance the productivity whereas sterical crowding lowers it. Among the different aluminoxane cocatalysts, methylaluminoxane is much more effective than ethyl- or isobutylaluminoxane. The catalyst shows a long lasting activity; even after more than 100 h polymerization time. The maximum activity is reached after 5 to 10 min (Fig. 3). This time seems to be needed the active site for form.

The concentration of catalyst and cocatalyst and the ethene pressure influence polymerization productivity. A near linear dependency of the activity on the concentration of ethene is observed. Polyethylenes produced by metallocene catalysts feature a molecular weight distribution of M$_W$/M$_n$ = 2 and 0.9 to 1.2 methyl groups per 1000 C atoms. Polyethylenes, catalyzed by different metallocenes, differ in their molecular weight by factors of more than 50. Catalysts possessing a substituent in the cyclopentadienyl or in the 2-position of an indenyl ligand give a higher molecular weight. Polyethylene produced with bis(pentamethylcyclopentadienyl) zirconium dichloride **10** possesses a molecular weight of 1.5 million. Similar results are observed with bis(tetrahydroindenyl) ansa-compounds **20, 25**. Mixing of different metallocene catalysts leads to bimodal distributions with MWD = 5–10. The molecular weight is easily lowered by increasing the temperature, raising the metallocene/ethene ratio, or by adding small amounts of hydrogen (0.1–2 mol%) [61].

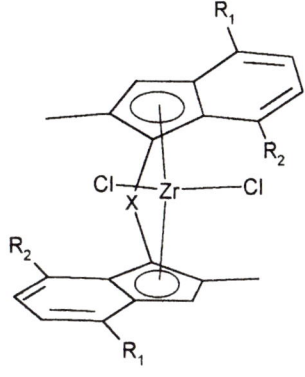

X = C₂H₄, Me₂Si

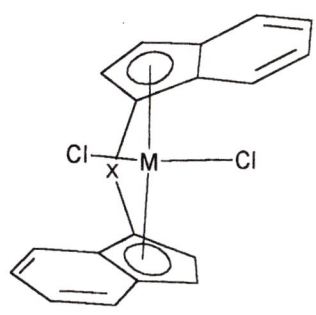

M = Zr,Hf X= C₂H₄, Me₂Si

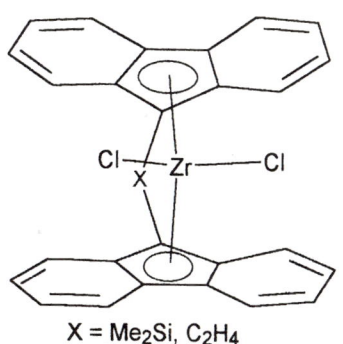

X = C₂H₄, R₁= R₂= Me
X = Me₂Si, R₁= R₂= Me
X = Me₂Si, R₁= Ph, R₂= H
X = Me₂Si, R₁= Naph, R₂= H

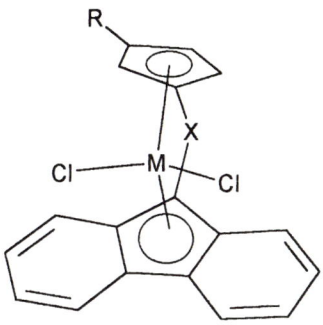

M = Zr, X= Me₂C, R= H
M = Hf, X= Me₂C, R= H
M = Zr, X= Ph₂C, R= H
M = Zr, X= Me₂C, R= Me
M = Zr, X= Me₂C, R= tBu

X = Me₂Si, C₂H₄

Fig. 2. Structures of selected metallocenes used for polymerization of olefins

Table 3. Homopolymerization of ethene at 30 °C, 2.5 bar ethene pressure, 6.25×10^{-6} mol/l metallocene concentration, molar ratio MAO/metallocene = 250. EAO = ethylaluminoxan, iBAO = triisobutylaluminoxan

No	Catalysts	Activity (kgPe/(molZr·h·[Et]))	mol.mass (g/mol)	Ref.
1	Cp$_2$ZrCl$_2$	60 900	620 000	56
2	Cp$_2$TiCl$_2$	34 200	400 000	
3	Cp$_2$HfCl$_2$	4 200[a]	700 000	
4	Cp$_2$TiMeCl	27 000[a]	440 000	53
5	Cp$_2$ZrMe$_2$	14 000[a]	730 000	53
6	Cp$_2$TiMe$_2$	1 200[a]	500 000	53
7	Cp$_2$HfMe$_2$	3 600[a]	550 000	53
8	Cp$_2$ZrCl$_2$/EAO	2 500[a]	550 000	57
9	Cp$_2$ZrCl$_2$/iBAO	5 400[a]	390 000	57
10	(C$_5$Me$_5$)$_2$ZrCl$_2$	1 300	1 500 000	
11	(Ind)$_2$ZrCl$_2$	45 000[a]	600 000	58
12	(neomenthylCp)$_2$ZrCl$_2$	12 200	1 000 000	56
13	(C$_5$Me$_4$Et)$_2$ZrCl$_2$	18 800	800 000	
14	[O(SiMe$_2$Cp)$_2$]ZrCl$_2$	57 800	930 000	
15	[O(SiMe$_2$tBuCp)$_2$]ZrCl$_2$	11 700	70 000	
16	[En(Ind)$_2$]ZrCl$_2$	41 100	140 000	56
17	[En(Ind)$_2$]HfCl$_2$	2 900	480 000	56
18	[En(Flu)$_2$]ZrCl$_2$	40 000[a]	–	
19	[En(2,4,7Me$_3$Ind)$_2$]ZrCl$_2$	78 000	190 000	
20	[En(IndH$_4$)$_2$]ZrCl$_2$	22 200	1 000 000	56
21	[Me$_2$Si(Ind)$_2$]ZrCl$_2$	36 900	260 000	56
22	[Ph$_2$Si(Ind)$_2$]ZrCl$_2$	20 200	320 000	56
23	[Bz$_2$Si(Ind)$_2$]ZrCl$_2$	12 200	350 000	56
24	[Me$_2$Si(2,4,7Me$_3$Ind)$_2$]ZrCl$_2$	111 900	250 000	
25	[Me$_2$Si(IndH$_4$)$_2$]ZrCl$_2$	30 200	900 000	
26	[Me$_2$Si(2Me4,6iPrInd)$_2$]ZrCl$_2$	18 600	730 000	
27	[Me$_2$Si(2Me4PhInd)$_2$]ZrCl$_2$	16 600	730 000	
28	[Me$_2$Si(2Me4,5benzoInd)]ZrCl$_2$	7 600	450 000	
29	[Ph$_2$C(Ind)(Cp)]ZrCl$_2$	3 330	18 000	
30	[Me$_2$C(Ind)(Cp)]ZrCl$_2$	1 550	25 000	56
31	[Me$_2$C(Ind)(3MeCp)]ZrCl$_2$	2 700	30 000	56
32	[Ph$_2$C(Flu)(Cp)]ZrCl$_2$	2 890	630 000	
33	[Me$_2$C(Flu)(Cp)]ZrCl$_2$	2 000	500 000	56
34	[Me$_2$C(Flu)(Cp)]HfCl$_2$	890	560 000	

[a] Calculated values from literature

At a reaction temperature of 10 °C, polyethylene is formed with a molecular weight of 1 500 000 g/mol, whereas at 50 °C this is reduced to 180 000 g/mol only to be lowered further to 90 000 g/mol at 90 °C. Raising the temperature to more than 100 °C yields α-olefins with an even number of carbon atoms.

With increasing zirconium concentration the molecular weight decreases nearly linearly. This leads to the conclusion that chain transfer occurs via a bimetallic mechanism.

The molecular weight can also be influenced by the addition of hydrogen (Fig. 4) [61]. In contrast to most heterogeneous catalysts, in this case only traces of hydrogen are needed to lower the molecular weight over a wide range. To

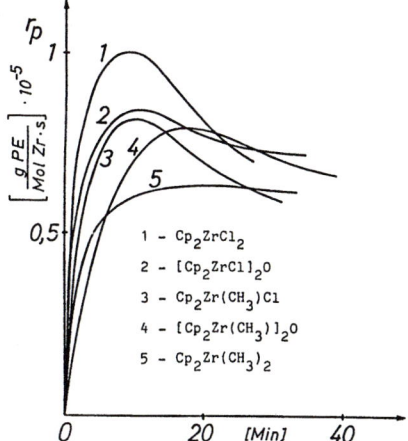

Fig. 3. Time dependence of the activity of various zirconocene/MAO catalysts for ethene polymerization at 30 °C

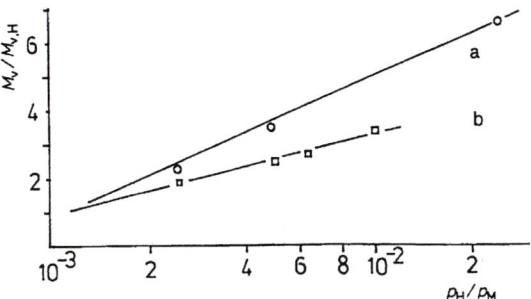

Fig. 4. Correlation of the ratio of viscosity average molecular weight M_v obtained without and with hydrogen $M_v/M_{v,H}$ to the ratio pH/pM of hydrogen and monomer (ethene) pressure in the feed. [Al] = 1.72×10^{-2} mol/l; line a-[Cp_2ZrCl_2] = 2×10^{-5} at 60 °C; line b-[Cp_2ZrCl_2] = 7.5×10^{-6} mol/l at 70 °C

reduce the molecular weight to about one half of that value obtained without hydrogen, only 7.5 vol.% H_2 is necessary to feed the reactor.

Table 4. Density and melting point of polyethylene samples with different molecular weights

M_w	M_w/M_n	Density of original material (g/cm³)	Density of pressed material (g/cm³)	Melting point (°C)
200 000	2.0	0.98	0.947	140.5
147 000	1.9	0.973	0.953	140.0
100 000	1.8	0.97		139.0

The density and melting points of the original material are high and decrease with molding steps and in samples with lower molecular weights (Table 4).

3.2 Copolymers

Metallocenes are highly useful for the copolymerization of ethene with other olefins. Propene, 1-butene, 1-pentene, 1-hexene, and 1-octene have been studied as comonomers, forming linear low density polyethylene (LLDPE). These copolymers have a great industrial potential and show a higher growth rate than the homopolymer. Due to the short branching from the incorporated α-olefin the copolymers show a lower melting point, lower crystallinity, and a lower density, making films more flexible and more processible. Applications of the copolymers can be found in packaging, in shrink films with a low steam permeation, in elastic films using high comonomer concentration, in cable coatings in the medical field because of the low amount of extractables, and in foams, elastic fibers, adhesives, etc. [3]. The comonomer is randomly distributed over the polymer chain. The amount of extractables is much lower than in polymer synthesized with Ziegler catalysts.

The copolymerization parameter r_1 which indicates how much faster an ethene is incorporated in the growing polymer chain than an α-olefin, when the last inserted monomer was an ethene unit, lies between 1 and 60 depending on the kind of comonomer and catalyst. The copolymerization parameter r_2 is the analogous ratio for the α-olefin. The product $r_1 \cdot r_2$ is important for the distribution of the comonomer and is close to unity when using C_2 symmetrie metallocenes, indicating a randomly distributed comonomer. It is less than unity with a more alternating structure for C_s-symmetric catalysts [62–65] (Table 5).

Under the same conditions, syndiospecific (C_s-symmetric) metallocenes are more effective for inserting α-olefins into an ethene-copolymer than isospecific working (C_2-symmetric) metallocenes, or unbridged metallocenes. In this case hafnocenes are more efficient than zirconocenes, too.

An interesting effect is observed for the polymerization with ethylene(bisindenyl) zirconium dichloride and some other metallocenes (Fig. 5). Although the activity of the homopolymerization of ethene is very high, it increases when copolymerizing with propene [66].

The measured data of the polymerization rate using a molar ratio of ethene/propene = 1:1 are four times higher than the calculated data. A clear increase in activity by the comonomer is observed. The results of the sequence analysis of the copolymer samples suggest no change in the mechanism of copolymerization. One explanation for this effect lies in the increase in the insertion rate due to an electronic influence of the comonomer.

The copolymerization of ethene with other olefins is effected by the variation of Al/Zr ratio, temperature and catalyst concentration [67, 68]. These variations

Table 5. Copolymerization parameters for ethene/α-olefin copolymerization by using different metallocene/MAO catalysts

Metallocene	Temp. (°C)	α-Olefin	r_1	r_2	$r_1 \cdot r_2$
Cp_2ZrMe_2	20	propene	31	0.005	0.25
$[En(Ind)_2]ZrCl_2$	50	propene	6.61	0.06	0.40
$[En(Ind)_2]ZrCl_2$	25	propene	6.26	0.11	0.69
$[Me_2C(Flu)(Cp)]ZrCl_2$	25	propene	1.3	0.20	0.26
Cp_2ZrCl_2	40	butene	55	0.017	0.93
Cp_2ZrCl_2	60	butene	65	0.013	0.85
Cp_2ZrCl_2	80	butene	85	0.010	0.85
$[En(Ind)_2]ZrCl_2$	50	butene	23.6	0.03	0.71
$[En(Ind)_2]ZrCl_2$	50	butene	8.5	0.07	0.59
Cp_2ZrMe_2	60	hexene	69	0.02	1.38
$[Me_2Si(Ind)_2]ZrCl_2$	60	hexene	25	0.016	0.40

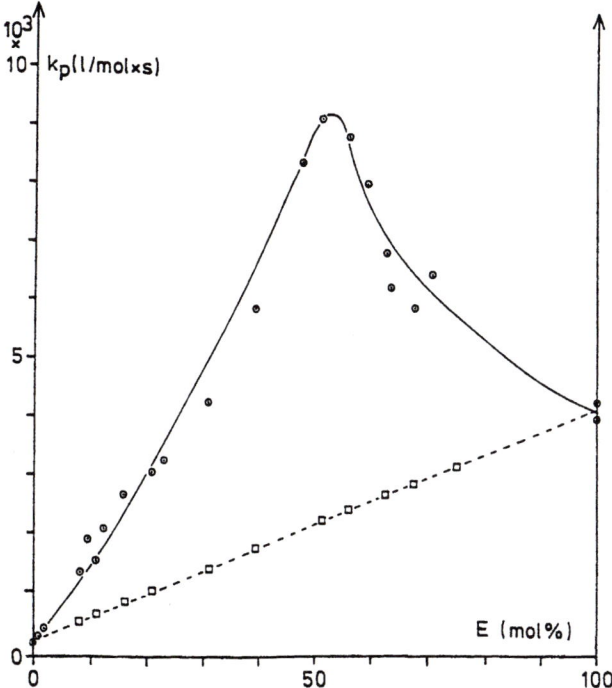

Fig. 5. Rate constant kp of the ethene/propene copolymerization as a function of the ethene concentration in the liquid phase at 37 °C; (- -) calculated; (-) measured

change the molecular weight and the ethene content. Higher temperatures increase the ethene content and lower the molecular weight.

Studies of ethene copolymerization with 1-butene using the Cp_2ZrCl_2/MAO catalyst indicated a decrease in the rate of polymerization with increasing comonomer concentration.

Of great industrial interest are the copolymers of ethene and propene with a molar ratio of 1/0.5, up to 1/2. These EP-polymers show elastic properties and, together with 2–5 wt% of dienes as third monomers, they are used as elastomers (EPDM). Since they have no double bonds in the backbone of the polymer, they are less sensitive to oxidation reactions. As dienes, ethylidenenorbornene, 1,4-hexadiene, and dicyclopentadiene are used. In most technical processes for the production of EP and EPDM rubber in the past, soluble or highly disposed vanadium components are used [69]. Similar elastomers can be obtained with metallocene/MAO catalysts by a much higher activity which are less colored [70–72]. The regiospecificity of the metallocene catalysts toward propene leads exclusively to the formation of head-to-tail enchainments. The ethylidenenorbornene polymerizes via vinyl polymerization of the cyclic double bond and the tendency to branching is low. The molecular weight distribution of about 2 is narrow [73].

At low temperatures the polymerization time to form one polymer chain is long enough to consume one monomer and add another. It thus becomes possible to synthesize block copolymers if the polymerization, especially catalyzed by hafnocenes, starts with propene and continues after the propene is nearly consumed with ethene.

3.3 Long Chain Branching

Branching, which is caused by the incorporation of long chain olefins into the growing polymer chain, is obtained with a new class of silyl bridged amidocyclopentadienyl titanium compounds (Fig. 6) [74–76].

These catalysts, in combination with MAO or borates, incorporate oligomers with vinyl endgroups which are formed during polymerization by β-hydrogen transfer. Thereby long chain branched polyolefins are formed, in contrast to linear polymers produced by other metallocenes. Copolymers of ethene with 1-octene are very flexible materials as long as the comonomer content is less than 10%. Upon reaching 20% the long branched polymers show elastic properties [77]. These catalysts are also able to copolymerize ethene and styrene [78–80].

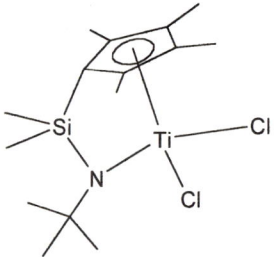

Fig. 6. Structure of dimethylsilylamido cyclopentadienyl titanium dichloride

4 Polypropene

4.1 Regio- and Stereospecificity

4.1.1 Microstructure of Polypropene

The microstructure of polypropene in terms of the enchainment of the monomer units and their configuration is determined by the regio- and stereospecificity of the insertion of the monomer. Depending on the orientation of the monomer during insertion into the transition-metal-polymeryl bond, primary (1,2-) and secondary (2,1-) insertions are possible (Fig. 7). Consecutive regiospecific insertion results in regioregular head to tail enchainment (1,3-branching) of monomer units while regioirregularities cause the formation of head to head (1,2-branching) and tail to tail (1,4-branching) structures.

Generally metallocenes favor consecutive primary insertions due to their bent sandwich structure. Secondary insertion also occurs to an extent determined by the structure of the metallocene used and the experimental setup (especially temperature and monomer concentration). Secondary insertions

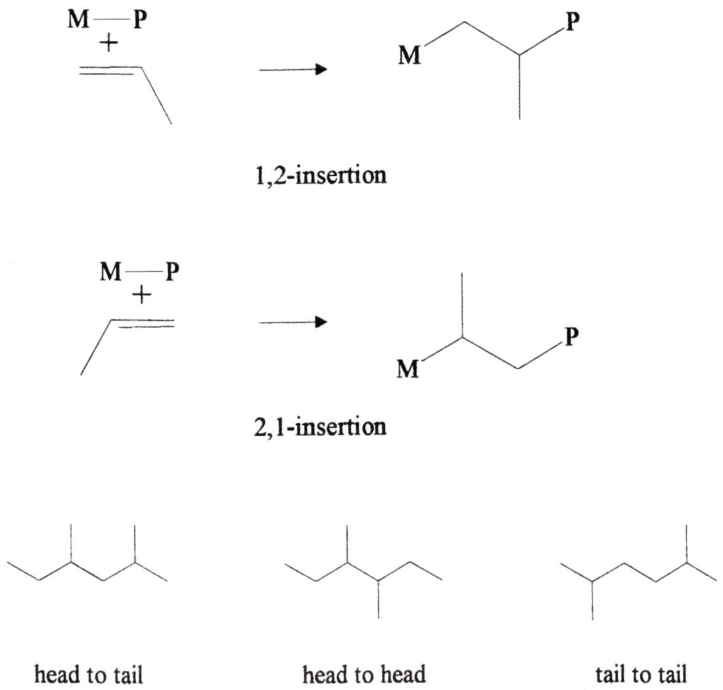

Fig. 7. Primary (1,2) and secondary (2,1) insertion in propene polymerization

cause an increased steric hindrance to the next primary insertion. The active center is blocked and therefore is regarded as a resting state of the catalysts [81]. The kinetic hindrance of chain propagation by another insertion favors chain termination and isomerization processes. One of the isomerization processes observed in metallocene catalysed polymerization of propene leads to the formation of 1,3-enchained monomer units (Fig. 8) [82–85]. The mechanism originally proposed to be of an elimination-isomerisation-addition type was recently discussed to involve transition metal mediated hydride shifts [86, 87].

Another type of steric isomerism observed in polypropene is related to the facts that propene is prochiral and polymers have pseudochiral centers at every tertiary carbon of the chain. The regularity of the configuration of successive pseudochiral centers determines the tacticity of the polymer. If the configuration of two neighboured pseudochiral centers is the same this "diad" is said to have a *meso* arrangement of the methyl groups (Fig. 9). If the pseudochiral centers are enantiomeric the diad is called *racemic*. A polymer containing only *meso* diads is called isotactic, while a polymer consisting of *racemic* diads only is named syndiotactic. Polypropene in which *meso* and *racemic* diads are randomly distributed is atactic (Fig. 10).

A single step of the polymerisation is analogous to a diastereoselective synthesis. Thus for achieving a certain level of chemical stereocontrol, chirality of the catalytically active species is necessary. In metallocene catalysis, chirality may be located at the transition metal itself, the ligand, or the growing polymer chain, e.g. the terminal monomer unit. Therefore two basic mechanisms of stereocontrol are possible [88, 89]: a) catalytic site control (also refered to as enantiomorphic site control), which is connected to chirality at the transition metal or the ligand, and b) chain end control which is caused by the chirality of the last inserted monomer unit. These two mechanisms cause microstructures (Fig. 11) which may be described by different statistics, while in the case of catalytic site control, errors are corrected due to the regime of the catalytic site (bernoullian statistics), chain end controlled propagation is not capable of doing so (markovian statistics).

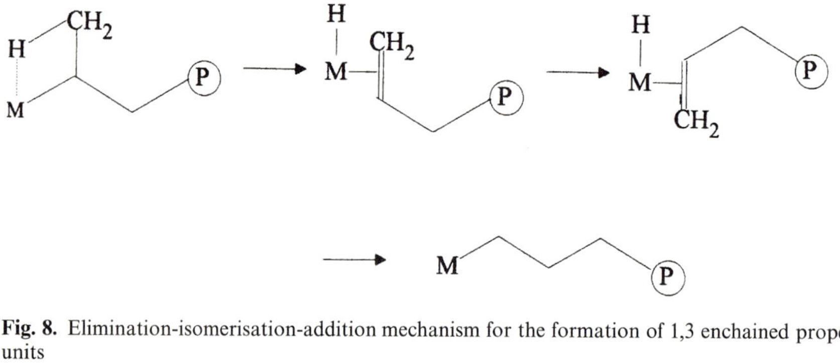

Fig. 8. Elimination-isomerisation-addition mechanism for the formation of 1,3 enchained propene units

racemic diad   meso diad

Fig. 9. Schematic drawing of the *racemic* and *meso* diad of poly(α-olefins)

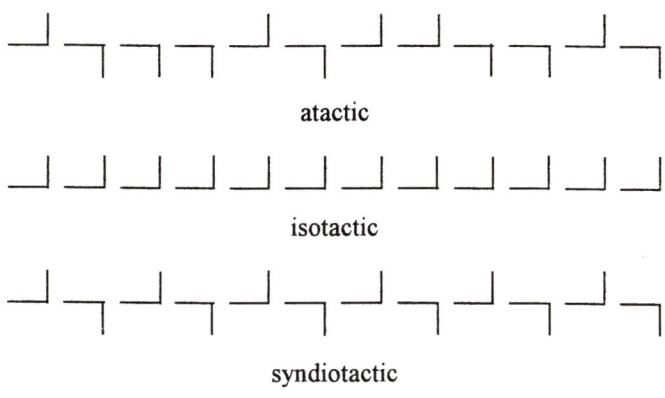

atactic

isotactic

syndiotactic

Fig. 10. Structures of polypropenes of different tacticities

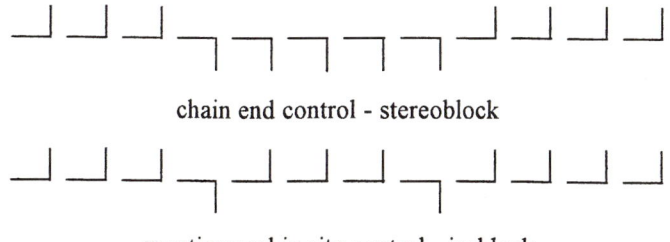

chain end control - stereoblock

enantiomorphic site control - isoblock

Fig. 11. Microstructures of polypropenes resulting from different types of stereocontrol during insertion

4.1.2 Symmetry and Stereospecificity

In the mid-1980s the first metallocene/MAO catalysts for the isotactic polymerization of propene were described. Ewen found Cp_2TiPh_2/MAO to produces isotactic polypropene at low temperatures by chain end control mechanism (stereoblock structure). When using a mixture of *racemic* and *meso* [En(Ind)$_2$] TiCl$_2$ in combination with MAO, he obtained a mixture of isotactic and atactic polypropene, the isotactic polymer having a microstructure in accordance with catalytic site control (isoblock structure). The use of pure racemic [En(Ind)$_2$] ZrCl$_2$ yielded for the first time pure isotactic polypropene formed by metallocene/MAO catalysts [90, 91]. These investigations were the beginning of rapid development in the area of metallocene catalysed polymerisation of propene which resulted in the invention of tailor-made metallocenes for different microstructures based on the mechanistic understanding of stereocontrol.

According to the structure of the metallocene, different microstructures are realized. Generally, among the rigid metallocenes, five different structures may be distinguished [92], and there are also metallocenes, which will be referred to later, having fluctuating structures.

4.1.2.1 C$_{2v}$ Symmetric Metallocenes

C$_{2v}$ symmetric metallocenes like Cp$_2$MCl$_2$ or [Me$_2$Si(Flu)$_2$]ZrCl$_2$ are achiral. The only stereocontrol observed is both chain end type and low, due to the fact that the chiral center of the terminal monomer unit of the growing chain is in β-position due to 1,2-insertion of the monomers. A significant influence on the tacticity is only observed at low temperatures (Table 6) and much more pronounced in case of titanocenes and hafnocenes due to their shorter M-Cα bonds bringing the chiral β-carbon closer to the active center [93–95].

Table 6. Chain end control by Cp$_2$M(2methylbutyl)$_2$/MAO at low temperatures determined as isotacticity index I.I. and sequence length of *meso* and *racemic* blocks

Temp (°C)	Zr/30	Zr/7	Zr/ − 20	Zr/ − 60	Ti/ − 35
mmmm%	0.052	0.085	0.106	0.140	0.430
n$_{iso}$	2.02	2.28	2.36	2.53	4.09
n$_m$	1.50	1.69	1.74	1.88	3.47
n$_r$	1.48	1.32	1.29	1.23	1.13

4.1.2.2 C$_2$ Symmetric Metallocenes

rac-Ethylenebis(indenyl)zirconium dichloride and rac-ethylenebis(4,5,7,8-tetrahydroindenyl)zirconium dichloride were the first chiral metallocenes investigated and found to produce isotactic polypropene. During the last ten years a lot of variations of these metallocenes have be published and patented, aiming for higher activities, molecular weights, tacticities and thereby higher melting points. Table 7 summarises some of the developments leading to catalysts which can produce polypropenes with properties comparable to the ones reached by using supported TiCl$_4$ catalysts [96–98].

Systematic investigation of bis(indenyl) zirconocenes showed that the main chain termination reaction is β-hydrogen transfer with the monomer [99, 100]. This reaction is very effectively suppressed by substituents (Me, Et) in position 2

Table 7. Comparison of the productivity, molecular weight, melting point and isotacticity obtained in polymerization experiments with various metallocene/MAO catalysts (bulk polymerization in 1 l liquid propene at 70 °C, Al/Zr ratio 15 000) showing the broad range of product properties) [96]

Metallocene	Productivity [kg^{PP}/(mmolZr × h)]	$M_w \times 10^{-3}$ [g/mol]	m.p. [°C]	Isotacticity [% mmmm]
[En(Ind)$_2$]ZrCl$_2$	188	24	132	78.5
[Me$_2$Si(Ind)$_2$]ZrCl$_2$	190	36	137	81.7
[Me$_2$Si(IndH$_4$)$_2$]ZrCl$_2$	48	24	141	84.5
[Me$_2$Si(2Me-Ind)$_2$]ZrCl$_2$	99	195	145	88.5
[Me$_2$Si(2Me-4iPr-Ind)$_2$]ZrCl$_2$	245	213	150	88.6
[Me$_2$Si(2,4Me$_2$-Cp)$_2$]ZrCl$_2$	97	31	149	89.2
[Me$_2$Si(2Me-4tBu-Cp)$_2$]ZrCl$_2$	10	19	155	94.3
[Me$_2$Si(2Me-4,5BenzInd)$_2$]ZrCl$_2$	403	330	146	88.7
[Me$_2$Si(2Me-4Ph-Ind)$_2$]ZrCl$_2$	755	729	157	95.2
[Me$_2$Ge(2Me-4Ph-Ind)$_2$]ZrCl$_2$	750	1135	158	–
[Me$_2$Si(2Me-4Naph-Ind)$_2$]ZrCl$_2$	875	920	161	99.1

of the indenyl-ring [99, 101, 102]. Substituents in position 4 also cause an enhancement in molecular weight by reducing 2,1-misinsertions which preferably result in chain termination by β-hydrogen elimination. Due to the fact that primary insertion is sterically hindered after a regioerror occurs and therefore the catalyst is in a resting state after a 2,1-insertion, suppression of this type of misinsertion also leads to enhanced activities. Using aromatic substituents in position 4 results in additional electronic effects. Thus the most active catalysts feature a methyl or ethyl group in position 2 and an aromatic group in position 4 of the indenyl rings.

Besides the bis(indenyl) ansa compounds, C_2 symmetric bridged bis(cyclopentadienyl) metallocenes of zirconium and hafnium were found to be able to produce isotactic polypropene (Table 8) [103]. The key for high isotacticity are substituents in positions 2,4,3' and 5' generating a surrounding of the transition metal similar to the one in bis(indenyl) metallocenes.

In this type of metallocenes the chirality is due to the chirality of the ligand and the two chlorines (e.g. the position of the growing chain and the

Table 8. Polymerization behavior of metallocenes based on bridged biscyclopentadienyl compounds. All polymerizations were performed at 30 °C in 500 ml toluene at 3 bar. Al/M = 10 000, [M] = 0.002 mmol, t = 2 h

Metallocene	Productivity [kgPP/(mmolZr × h)]	$M_w \times 10^{-3}$ [g/mol]	m.p. [°C]	Isotacticity [% mmmm]
[Me$_2$Si(2,3,5Me$_3$Cp)$_2$]ZrCl$_2$	1.6	134	162	97.7
[Me$_2$Si(2,4Me$_2$Cp)$_2$]ZrCl$_2$	11.1	87	160	97.1
[Me$_2$Si(3tBuCp)$_2$]ZrCl$_2$	0.3	10	149	93.4
[Me$_2$Si(3MeCp)$_2$]ZrCl$_2$	16.3	14	148	92.5
[Me$_2$Si(2,3,5Me$_3$Cp)$_2$]HfCl$_2$	0.30	256	163	98.7
[Me$_2$Si(2,4Me$_2$Cp)$_2$]HfCl$_2$	0.10	139	162	98.5
[Me$_2$Si(3tBuCp)$_2$]HfCl$_2$	0.03	17	157	–
[Me$_2$Si(3MeCp)$_2$]HfCl$_2$	1.61	67	148	–

coordinating monomer) are homotopic. According to a model of Pino et al. [104–107] the conformation of the growing polymer chain is determined by the structure of the incoming monomer and is forced into a distinct orientation by steric interactions of its side chain with the polymer chain (Fig. 12) [108–113] (the relative topicity of this reaction was found to be *like*). Due to the C_2-symmetry and the homotopicity of the coordination vacancies isotactic polymer is produced (Fig. 13).

4.1.2.3 C_s-symmetric Bridged Metallocenes

In 1988 Ewen and Razavi developed a catalyst for the syndiotactic polymerisation of propene based on Cs symmetric metallocenes (Table 9) [114–116].

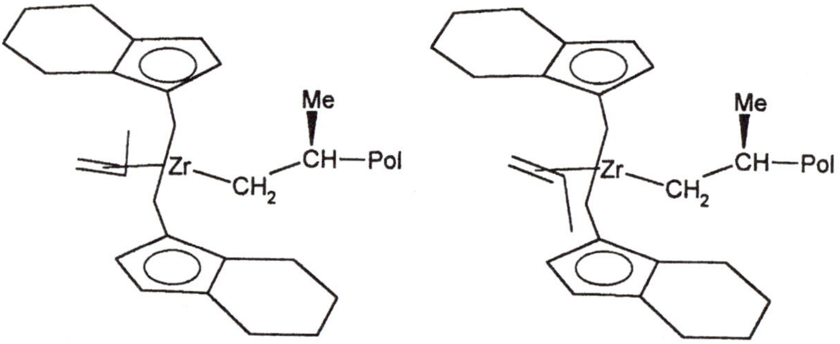

Fig. 12. Origin of the stereospecificity of C_2 symmetric bis(indenyl) zirconocene catalysts. The orientation shown *on the right* is favored over the one *shown left* due to non-bonding interaction of the approaching monomer and the ligand

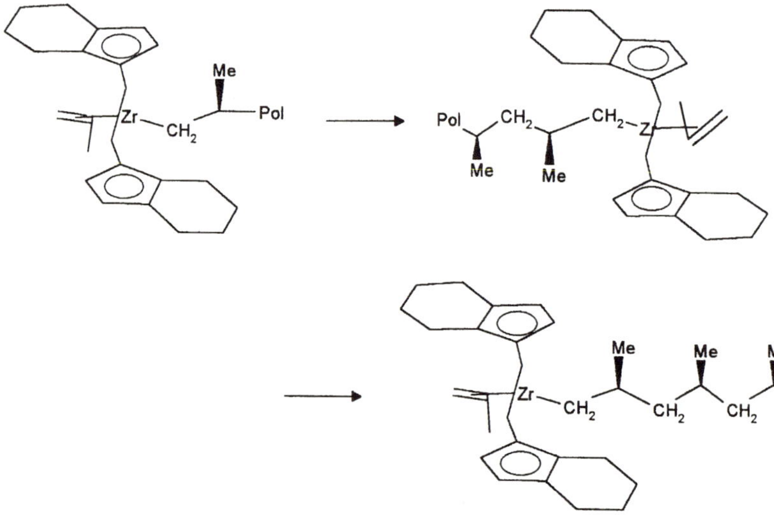

Fig. 13. Mechanism of the isotactic polymerization of propene using an alkylzirconocenium ion generated from a C_2 symmetric bis(indenyl) zirconocene

Table 9. Syndiotactic polypropenes prepared by different metallocene catalysts. Polymerisations were carried out at 60 °C in 1 l of liquid propene

Metallocene	Productivity [kgPP/(gM h)]	M_w [kg/mol]	m.p. [°C]	Syndiotacticity [rrrr %]
$[Me_2C(Flu)(Cp)]ZrCl_2$	180	90		0.82
$[Me_2C(Flu)(Cp)]HfCl_2$	3	778		0.73
$[En(Flu)(Cp)]ZrCl_2$	50	171	111	0.71
$[Ph_2C(Flu)(Cp)]ZrCl_2$	3138	478	133	0.87
$[Ph_2C(Flu)(Cp)]HfCl_2$	28	1950	102	0.74

From these prochiral metallocenes, chiral metallocenium ions can be produced in which chirality is centred at the transition metal itself. Due to the flipping of the polymer chain the metallocene alternates between the two enantiomeric configurations (Fig. 14) and produces a syndiotactic polymer [114–118].

4.1.2.4 C_1-symmetric Bridged Metallocenes

Variation of C_s symmetric metallocenes leads to C_1 symmetric ones (Fig. 15). If a methyl group is introduced at position 3 of the cyclopentadienyl ring, stereospecificity is disturbed at one of the reaction sites so every second insertion is random. A hemiisotactic polymer is produced [119–121]. If steric hindrance is bigger (for example a *tert*-butyl group is introduced instead of the methyl group) stereo selectivity is inverted and the metallocene catalyses the production of isotactic polymers (Table 10) [117, 118, 122–125].

4.1.2.5 Chain Migratory Insertion, Chain Stationary Insertion, and Site Isomerization

The microstructures described above correlate to chain migratory insertion. While in the case of a C_2 or C_{2v} symmetric metallocene, due to the homotopic nature of the potentially active sites, chain stationary insertion or migratory insertion followed by site isomerization would result in the same microstructure as chain migratory insertion, in the case of C_s symmetric catalysts they result in isotactic blocks.

C_1 symmetric metallocenes are able to produce new microstructures if consecutive insertions take place on the same active site in addition to chain migratory insertion. Polypropenes containing blocks of atactic and isotactic sequences are produced, the block lengths depending on the rate of chain stationary insertion or site isomerization vs chain migratory insertion [126–133]. Thus hemiisotactic polypropene represents a special case, having a chain stationary/chain migratory ratio of 1:1.

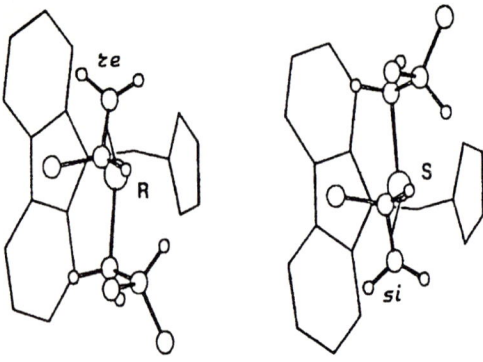

Fig. 14. Origin of syndiotacticity in polypropene produced by [Me$_2$C(Flu)(Cp)]ZrCl$_2$: the polymer chain allows only a distinct orientation of the monomer and the chirality is located at the transition metal center, and therefore the catalyst alternates between the two enantiomeric forms

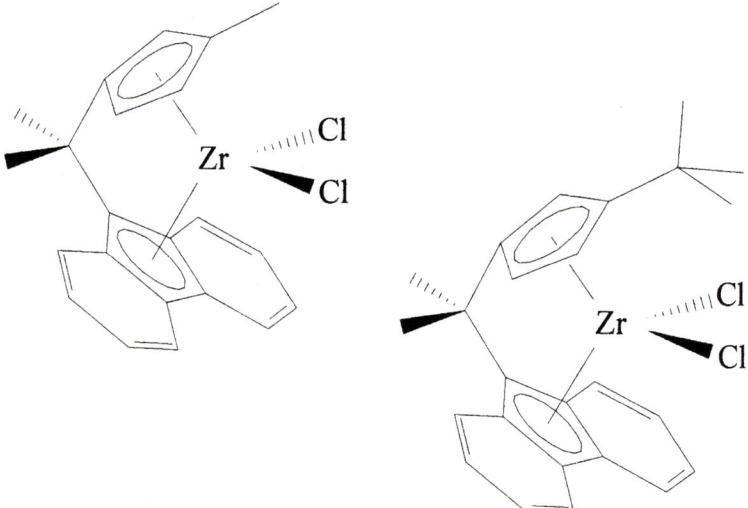

Fig. 15. Structures of [Me$_2$C(Flu)(3MeCP)]ZrCl$_2$ and [Me$_2$C(Flu)(3tBuCp)]ZrCl$_2$

Table 10. Changing the polymerization behavior from syndiotactic via hemiisotactic to isotactic by introducing substituents in Cs symmetric metallocene. Polymerizations were carried out at 60 °C in 1 l of liquid propene

Metallocene	Activity [kg/gCat]	Tm [°C]	M$_w$ [kg/mol]	mmmm %	rrrr%
[Me$_2$C(Flu)(Cp)]ZrCl$_2$	180	136	90	0	84
[Me$_2$C(Flu)(3MeCp)]ZrCl$_2$	25	−15	36	23.6	17.83
[Me$_2$(Flu)(3tBuCp)]ZrCl$_2$	48	127	62	77.5	0.6

4.1.2.6 Oscillating Metallocenes

Polymers of a similar microstructure are obtained if unbridged substituted metallocenes having a significant rotational isomerisation barrier are used as catalyst (Fig. 16). Early attempts concentrated on substituted cyclopentadienyl and indenyl compounds [134–138].

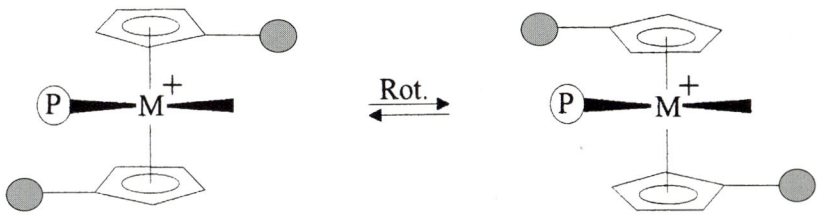

Fig. 16. Oscillating metallocene: by rotation of the cyclopentadienylrings the metallocenes epimerizes

Most recent efforts by Coates and Waymouth have shown that 2-phenylindenyl-groups are well suited for this purpose [139]. They oscillated between the enantiomeric and meso arrangements giving rise to a stereoblock polypropene containing atactic (produced by the meso rotamer) and isotactic (produced by the chiral rotamer) sequences. The block length was strongly dependent on the temperature.

4.2 Oligomerisation and Mechanism

Based on the Cossee-Arlman mechanism, Pino et al. have proposed a model for the explanation of the origin of stereoselectivity. The metallocene forces the polymer chain into a distinct arrangement which in turn determines the stereochemistry of the approaching monomer (Fig. 17). This model is stressed by experimental observations in metallocene catalyzed oligomerisation.

One way of controlling the molecular weight of polymers produced by Ziegler-Natta catalysts is through the addition of hydrogen which causes chain termination by transfer. At high hydrogen concentrations, oligomers are produced. If an enantiomerically pure metallocene like (S)-[En(IndH$_4$)]ZrCl$_2$ is used as catalyst, optically active oligomers are formed. From optical rotation of these oligomers the topicity of insertion may be determined as well as the enantiomeric excess.

The first insertion into a Zr-H bond was found to occur with relative topicity *unlike*: in accordance with the fact that no polymer chain is present, the orientation of the inserting monomer being determined by the ligand framework. The second insertion as well as the following ones occur with the relative topicity *like*: the polymer chain inverts the preferred mode of insertion [140–143]. The enantiomeric excess at 30 °C calculated for the first insertion is

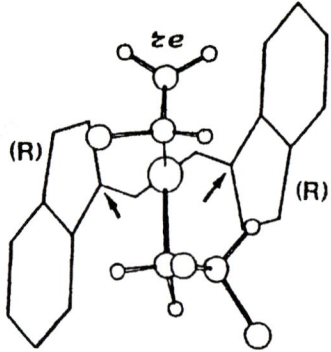

Fig. 17. Origin of stereospecificity in propene polymerization by ethylenebis(indenyl) zirconocendichloride: the growing polymer chain forces the inserting monomer into an orientation resulting in the relative topicity *like* for insertion

Table 11. Comparison of the temperature dependence of the enantiomeric excess (calculated from mmmm 0.2) obtained in bulk polymerization of propene to the ee observed in the trimer fraction produced at [propene] → 0 by (S)-[En(IndH$_4$)]ZrCl$_2$/MAO

Temp (°C)	mmmm	b	1 − b	ee%
		Polymerization		
0	0.95	0.989	0.011	97.8
30	0.94	0.988	0.012	97.6
60	0.89	0.977	0.023	95.4
		Oligomerizaton		
20	0.89	0.977	0.023	95.3
30	0.49	0.867	0.133	73.4
40	0.25	0.757	0.243	51.3
50	0.10	0.619	0.381	23.8
60	0.07	0.555	0.445	10.9
70	0.06	0.513	0.487	2.5

about 94%, while for the second insertion it is 97.8% which is in accordance with an enantiomeric excess of 98% calculated from the %mmmm of a polymer produced in bulk polymerization (Table 11).

If polymerisation is carried out not in bulk at lower monomer concentration, enantiomeric excess and thereby tacticity decreases [144, 145]. Leclerc and Brintzinger have proven that epimerization of the last monomer unit of the growing chain by hydrogen migration from the Me-group is responsible for about 80% of all stereodefects observed at low monomer concentrations [146]. A mechanism proposed by Busico et al. is of an elimination-isomerisation-readdition type similar to the ones discussed for 1,3-insertions. Thus as for all types of isomerisations observed in metallocene catalyzed polymerisation, conditions which favor β-agostic stabilization of the active species may lead to lower enantiomeric excess. This also may be concluded from the observation that with increasing temperature at low monomer concentration the enantiomeric excess is dramatically lowered, resulting in the production of atactic oligomers at [propene] $\rightarrow 0$ mol/l and temperatures of 50–60 °C [147–149].

4.3 Polymer Properties

4.3.1 Atactic Polypropene

So far only low molecular weight atactic polypropene (aPP) waxes with broad molecular weight distributions, produced by solvent extraction from iPP prepared by Ziegler-Natta catalysts and used as additive in bitumen or adhesives have been known. Using metallocene catalysts, aPPs covering the whole range of molecular weights of technical interest with narrow molecular weight distributions can be produced [150]. The main characteristics of high molecular weight aPP produced by $[Me_2Si(Flu)_2]ZrCl_2$ are low density, high transparency, softness, low modulus, and high elongation caused by the totally amorphous state of the polymer (Table 12). High molecular weight aPP has potential applications in blends with other polyolefins, upgrading transparency, softness, elastic recovery and elongation.

Table 12. Comparison of low molecular weight atactic polypropene from solvent extraction of a polymer produced using a conventional $TiCl_4/MgCl_2$ catalyst with two high molecular weight atactic polypropenes of different molecular weight prepared by $[Me_2Si(Flu)_2]ZrCl_2MAO$. For comparison typical data of an isotactic polypropene prepared by a conventional catalyst are also given [150]

	Low Mw aPP	High M_waPP		iPP
Intrinsic viscosity (dl/g)	0.4	1.4	2.7	2.3
Melt index (dg/min)	670	7	0.1	1.8
M_w*10^{-3} (g/mol)	29	200	490	520
M_w/M_n	6	3.3	2.3	8
Density (g/cm^3)	0.8626	0.8606	0.8550	0.9031
Shore A hardness (°shore)	67	50	55	–
Haze (%)	58	20	18	85
Modulus (MPa)	10	8	5	1650
Strength at break (MPa)	1	1	2	20
Elongation at break (%)	110	1400	2000	300

4.3.2 Isotactic Polypropene

The properties and melting point of isotactic polypropenes prepared by metallocene catalysts are determined by the amount of irregularities (stereo- and regioerorrors) randomly distributed along the polymer chain. Thus the term stereospecificity does not refer to extractable aPP as for conventional PP always

Table 13. Comparison of isotactic polypropenes pepared by different metallocene/MAO catalysts- [En(IndH$_4$)$_2$ZrCl$_2$ (I), [Me$_2$Si(4,5BenzInd)$_2$]ZrCl$_2$ (II), [Me$_2$Si(4,6iPrInd)$_2$]ZrCl$_2$ (III)- at 70 °C in a 1 l bulk polymerization at Al/Zr = 15 000 to conventional isotactic PP prepared by a TiCl$_4$/MgCl$_2$ catalyst (IV) [156]

	(I)	(II)	(III)	(IV)
Melting point [°C]	139	151	160	162
M_w/M_n	2.2	2.3	2.5	5.8
Modulus [N/mm^2]	1060	1440	1620	1190
Hardness [N/mm^2]	59	78	86	76
Impact resistance Izod [mJ/mm^2]	128	86	100	103
Light transmission rate [% 1-mm plate]	56	44	35	34
Melt flow rate [°/min]	2	2	2	2

having a melting point of 160–165 °C. Metallocene catalysts depending on their substitution pattern can give a wide range of homopolymers having melting points between 125 and 165 °C (Table 13).

The molecular weight distribution of these iPPs (Mw/Mn = 2–2.5) is still lower than that of conventional controlled rheology grade obtained by peroxide degradation (Mw/Mn = 3–4) inducing good performance in procession by thin wall molding or fiber extraction. For applications demanding broader molecular weight distributions, two or more metallocenes may be combined to give a tailor-made molecular weight distribution. Compared with conventional iPP grades, metallocene products show enhanced mechanical strength which can be improved by tailoring the molecular weight distribution.

The low melting points obtained with some metallocene catalysts, even at high pentad isotacticities, are caused by 2,1- and 1,3-misinsertions [151, 152]. Low melting point polymers with conventional catalysts are obtained by copolymerization with small amounts of ethene. A comparison of a conventionally produced polymer having the same melt flow rate and low melting point as a metallocene copolymer and a metallocene homopolymer shows the enhanced stiffness and transparency of the metallocene product (Table 14). The most important feature is the low amount of extractables, allowing the use of polypropenes for food wrapping and applications at cooking temperature.

The excellent performance of metallocenes in copolymerizations also offer improvements in impact copolymers. In the wide variety of properties of impact copolymers, the stiffness of the material is determined by the matrix material, while the impact resistance largely depends on the elastomeric phase. While conventional catalysts show some inhomogeneities in the ethene/propene rubber phase due to crystalline ethene rich sequences, the more homogenous comonomer distribution obtained with metallocene catalysts results in a totally amorphous phase [153].

Using highly stereoselective metallocenes, highly crystalline, stiff polypropene types are produced. These polymers exhibit a stiffness 25–30% above that of conventional polypropenes, resembling that of polypropenes filled with

Table 14. Comparison of low melting point polymers obtained by copolymerization using a conventional catalyst, a metallocene catalyst and by homopolymerization using a metallocene catalyst. [153]

	Conv. copolymer	Metallocene copolymer	Metallocene homopolymer
Melting point [°C]	141	140	142
Modulus [N/mm^2]	620	940	1120
Hardness [N/mm^2]	41	59	65
Impact resistance Izod [mJ/mm^2]	23.1	11.3	7.3
Light transmission [% 1-mm plate]	57	65	48
Extractables [% hexane, 69 °C]	7.9	1.1	0.7

Table 15. Mechanical properties of highly stiff polypropenes compared to normal materials [154]

	pp-homopolymer		pp-blockcopolymer	
Stiffness	high	normal	high	normal
Application	thermoforming		thin wall injection molding	
MFR 230/2.16	3.5	3.5	58	45
e-modulus (MPa)	1950	1500	1550	1350
Tensile strenghth (MPa)	38	35	29	26
Izod impact strength at 23 °C/ − 30 °C (kJ/m)	85/10	80/15	70/25	110/30
HDT A/B (°C)	60/108	55/85	57/98	53/90

talcum or other minerals (Table 15) [154]. Packages made from these polypropenes may have reduced thickness of the walls, are easier recycled, show enhanced impact strength, heat resistance, lower density and less aging.

Metallocenes are also interesting for the production of new iPP waxes for use as pigment dispersants, toner or lacquer surfaces [155]. The molecular weight of about 10 000 to 70 000 g/mol combined with melting points between 140 and 160 °C is easily obtained by the choice of the metallocene, (Table 16) while, with conventional catalysts, hydrogen or polymer degradation are used to control the molecular weight and choice of the donor or addition of a comonomer to control the melting point. As can be seen in Table 8, metallocene catalysts offer property combinations not accessible with conventional systems; thus, for example, the vinyl endgroups may be utilized for functionalization while, with conventional catalysts, only saturated endgroups are formed due to the high amount of hydrogen used for molecular weight regulation. Process simplification stems from the use of metallocene catalysts. With conventional catalysts the reactor has either to be run under non-optimal conditions (high temperature, high hydrogen pressure ⇒ lower productivity) with decreased output due to the difficulty of condensing hydrogen/propene feeds with large amounts of hydrogen and the low heat removal capacity of these mixtures or, if run at optimum conditions, polymer degradation with expensive peroxides has to be used for metallocenes to avoid these difficulties.

Table 16. Isotactic polypropene waxes prepared by [Me$_2$Si(2Me-4tBu-Cp)$_2$]ZrCl$_2$/MAO (I), [Me$_2$Si(IndH$_4$)$_2$]ZrCl$_2$/MAO (II) compared to waxes produced by conventional catalysts via molecular weight regulation by hydrogen (III) or visbreaking of PP random copolymers (IV) or homopolymers (V) [155]

	(I)	(II)	(III)	(IV)	(V)
M_w*10^{-3} [g/mol]	68	50	40	44	36
M_w/M_n	1.8	2.0	3.8	2.2	1.9
C_2/C_4 [%]	–	–	–	4.0/2.4	–
m.p. [°C]	163	133	159	133	155
Crystallinity [%]	69	50	60	30	59
Isotacticity [mmmm %]	96	85	91	80	91
Double bonds/chain	0.5–1	1	0	4	5
Misinsertions/1000C	1.7 1.3	4.7 1.3		19 C2	n.d.
	0.2 2.1	0.3 2.1	0.3 2.1	7 C4	
Melt viscosity (200 °C) [mm^2/s]	2800	2521	900	1684	1040
Hardness [bar]	2000	874	1800	423	1870
Yellowness index	0.5	0.8	1	5–6	1–2
Dropping point [°C]	168	147	160	149	169
Congealing temp. [°C]	n.d.	113	117	102	124

4.3.3 Syndiotactic Polypropene

Syndiotactic polypropene produced by metallocene catalysts shows a higher level of irregularities than isotactic ones. Comparing samples of the same degree of tacticity, (Table 17) the syndiotactic polymer exhibits a lower melting point, lower density (strongly) depending on the tacticity, ranging from 0.87 to 0.89 g/cm^3), lower crystallinity, and a lower crystallisation rate [156]. The small crystal size in syndiotactic polypropene causes a higher clarity of the material but is also responsible for its inferior gas barrier properties preventing applications in food packaging. However the resistance against radiation allows medical applications. Other advantages of sPP are the higher viscous and elastic moduli at higher shear rates and its outstanding impact strength which disappears at low temperatures due to the independence of the glass transition temperature on the tacticity.

Commercial product of syndiotactic polypropene utilizes a silica supported metallocene in a bulk suspension process at 50–70 °C and a pressure of 30 kg/cm^2 [157].

The combination of flexibility, clarity and tensile set and low heat seal temperatures (Table 18) enables syndiotactic polypropenes to be applied instead of PVC, EVA and LLDPE in films, foils and extruder products.

4.3.4 Elastomeric Polypropene

Elastomeric polypropenes (Table 19) of two different types may be prepared using metallocene catalysts: 1) polypropenes being elastomeric due to a high

Table 17. Comparison of isotactic and syndiotactic polypropene produced by metallocene catalysts

	$[Me_2Si(Ind)_2]ZrCl_2$	$[Me_2C(Flu)(Cp)]ZrCl_2$
Tacticity	iso	syndio
mmmm%/rrrr%	83.1	83.6
niso/nsyn	33	25
m.p. [°C]	138.4	133.2
Crystallinity DSC [%]	41.6	27.2
MFI 230/5	16.4	21.1
Density	0.899	0.885

Table 18. Mechanical properties of syndiotactic polypropene compared to a random ethene/propene copolymer [156]

	Syndiotactic PP	Random copolymer
Melt index (g/10min.)	3.0	8.0
Haze (%)	1.7	3.0
e-modulus (MPa)	61	60
Impact strenghth (kJ/m^2)	80	90
Heat seal temp. at 120 g/cm (°C)	135	138
Heat seal temp. at 200 g/cm (°C)	137	143

Table 19. Properties of elastomeric polypropenes prepared by $[MeHC(Ind)(C_5Me_4)]TiCl_2$ (1), $[Me_2C(Ind)(Cp)]HfCl_2$ (2), $[Me_2C(Ind)(Cp)]ZrCl_2$ (3) and $(2PhInd)_2ZrCl_2$ (4) [128, 132]

Catalysts	1	2	3	3	4
M$_w$ (kg/mol)	127	30	50	380	889
mmmm%	40	38	54	52	28
Tm		47/61	54/93	53/84	125–145
H$_f$ (J/g)	15		40	35	0.4
Crystallinity % From H$_f$/209	6.7	7.2	19.1	16.7	0.2
strain to break %	525	200	500	800	1210
Elastic recovery					
after 100% strain	93		92	95	
after 200% strain	91		90	93	
after break	86		86	84	

content of 1,3-enchainments; and 2) polypropenes having a stereoblock structure prepared using oscillating or C_1 symmetric metallocenes.

4.4 Heterogenization and Polymerization in the Presence of Fillers

Metallocene catalysts which are to be used as drop-in catalysts in existing plants for polyolefin production have to be heterogenized due to the fact that current technology is based on gas phase and slurry processes. Thus the metallocenes are to be fixed on a carrier. Carriers may be divided into three groups: 1) metals have been used as fillers; 2) inorganics like silica, aluminia, zeoliths or $MgCl_2$ [158–165]; and 3) organic materials like cyclodextrins [166], starch (as a filler) [167] and polymers (polystyrenes, polyamides) have been used to support either the metallocene or the cocatalyst.

Looking at the preparation of supported metallocenes, synthesis of the metallocene on the carrier is found as well as fixing a metallocene either via functionality at the ligand or by direct reaction with the carrier, in both cases followed by activation with MAO or trialkylaluminium, but more common is heterogenization of the cocatalyst prior to mixing the modified carrier with the metallocene and activation by trialkylaluminium.

4.4.1 Synthesis of a Metallocene on a Carrier

Soga et al. have reported the synthesis of ansa zirconocenes on silica by synthesizing a precursor of the bridge anchored using SiO_2–OH groups on the surface of silica (Table 20) [168].

Activation of these catalysts is done using either MAO or triisobutylaluminium. In all cases the polymers obtained are isotactic despite the fact that synthesis of fixed bis(indenyl) metallocenes may result in inseparable meso and racemic diastereomers and bridged bis(fluorenyl) metallocenes are not chiral.

4.4.2 Attaching a Metallocene to a Carrier

A procedure similar to the one described above starts with a metallocene carrying an additional functionality at the ligand (Fig. 18) either on the cyclopentadienyl rings or at the bridge which may be used to fix it [169–173].

These metallocenes may be fixed on silica (probably after modification of the SiO_2–OH groups) or other inorganic carriers as well as on polymeric materials.

Metallocenes may also be fixed on silica or alumina by direct reaction of the two components. Marks has shown that reaction of dimethyl metallocenes with alumina results in the formation of an active catalyst [174]. Others have investigated the direct reaction of metallocene dichlorides with silica (and alumina) and faced the problem of metallocene decomposition [175–177]. Nevertheless an active species is formed which produces isotactic polypropene but with rather low activities.

4.4.3 Fixing a Metallocene on Supported MAO

Two approaches have been followed to generate a supported methylaluminoxane: 1) the reaction of a carrier containing hydroxyl groups (starch,

Table 20. Polymerization behavior of silica-fixed metallocene catalysts at 40 °C in toluene

Catalyst	Cocatalyst	Tm iso PP (°C)
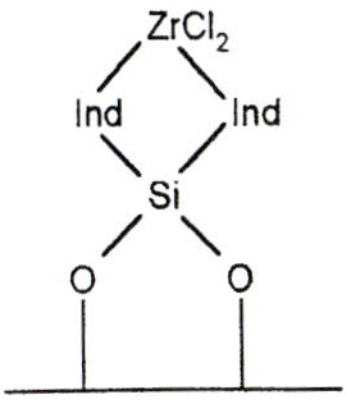 Ind–Si(ZrCl$_2$)–Ind, Si bridged to silica via two O	MAO	162/156
	TIBA	162/159
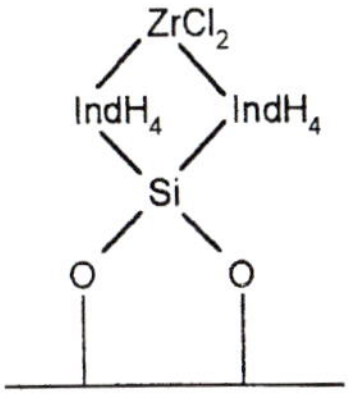 IndH$_4$–Si(ZrCl$_2$)–IndH$_4$, Si bridged to silica via two O	MAO	138
	TIBA	137
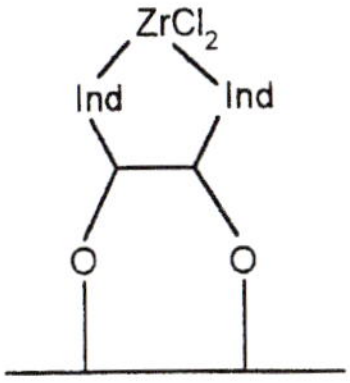 Ind–(ZrCl$_2$)–Ind, ethylene-bridged to silica via two O	MAO	160/150
	TIBA	158
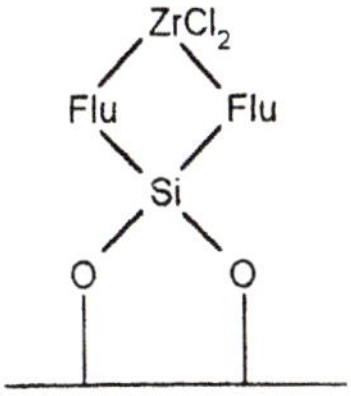 Flu–Si(ZrCl$_2$)–Flu, Si bridged to silica via two O	TIBA	160

silica) with trimethylaluminium; and 2) fixing MAO itself by reaction with OH groups of the carrier. In both cases heterogenization of the cocatalyst is followed by reaction of the heterogeneous MAO with a metallocene dichloride to generate a metallocene bond to the supported MAO. These catalysts are usually activated by trialkylaluminiums or additional small amounts of MAO. Metallocenes fixed on supported MAO exhibit similar behavior in polymerization as do their homogenous analogues (Table 21). Transfer of knowledge about catalysts design gathered in homogenous processes is possible. Therefore these techniques are most widely used to fix metallocenes onto a carrier.

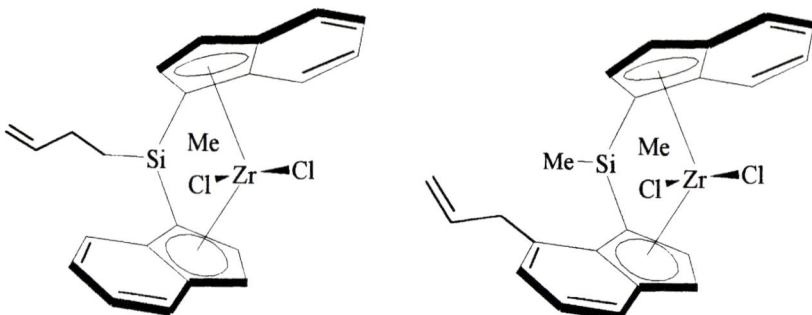

Fig. 18. Possible structures for metallocenes containing a functional group in the ligand either at the bridge or at the indenyl (cyclopentadienyl) group, making it possible to attach the metallocene to a carrier or even perform a polymerization using the metallocene as a comonomer

Table 21. Comparison of metallocenes in homogenous phase and supported on silica fixed MAO at 40 °C. (I) $[En(IndH_4)]_2ZrCl_2$, (II) $[Me_2C(Flu)(Cp)]ZrCl_2$, (III) Cp_2ZrCl_2

Catalyst	Cocatalysts	Activity (kg/molZr·h)	Tm (°C)	M_n (kg/mol)	M_w/M_n	mmmm% or rrrr%
(I)/homogenous	MAO 3 mmol	2070	111	3.3	1.9	71
(I)/MAO/SiO2	TMA 1 mmol	313	126	2.2	1.8	–
(I)/MAO/SiO2	TEA 1 mmol	77	140	5.3	2.5	90
(I)/MAO/SiO2	TIBA 1 mmol	556	136	14.2	2.0	–
(I)/MAO/SiO2	TIBA 0.5 mmol	> 1500	128	4.7	3.1	–
(I)/MAO/SiO2	TIBA 2 mmol	382	105	6.6	1.8	69
(II)/homogenous	MAO 13 mmol	758	123	39.3	1.8	77
(II)/MAO/SiO2	TIBA 2 mmol	141	133	45.2	1.9	83
(III)/homogenous	MAO 10 mmol	132	–	0.3	–	–
(III)/MAO/SiO2	TIBA 0.4 mmol	99	–	1.8	–	–

5 Cycloolefin Polymerization

5.1 Homopolymers

Strained cyclic olefins like cyclobutene, cyclopentene, and norbonene can be used as monomers and comonomers in a wide variety of polymers. Generally they can be polymerized by ring opening polymerization (ROMP) featuring elastomeric materials [178–180] or by double bond opening (vinyl polymerization). Homopolymerization of cyclic olefins by double bond opening is achieved by several transition metal catalysts, namely Pd catalysts [181–187] and metallocene catalysts (Fig. 19).

The polymers feature two chiral centers per monomer unit and therefore are ditactic. While polymers produced by achiral Pd catalysts seem to be atactic, using chiral metallocene catalysts highly tactic crystalline materials can be produced, featuring extraordinary high melting points (in some cases above the decomposition temperature) and extreme chemical resistance.

The microstructures of these polymers have been investigated using oligomers as models. Norbornene was shown to polymerize via *cis exo* insertion [188, 189], while in the case of cyclopentene quite unusual *cis* and *trans* 1,3 enchainment of the monomer units is observed [190–192].

Cyclopentene has been polymerized by several metallocene catalysts (Table 22). For all of them the homopolymers were found to contain no 1,2-enchainments. While for the formation of *cis*-1,3 enchainments a mechanism similar to that proposed for the formation of 1,3-enchainments in polypropene is reasonable, there is no plausible explanation for the formation of trans structures.

Investigation of the polymers produced at high catalysts concentrations shows their rather low molecular weights which goes along with melting over a broad range from 150–350 °C. At low catalyst concentrations crystalline polymers not melting up to about 400 °C may be produced if a chiral metallocene is used. The glass transition temperature of poly(cyclopentene) is about 65 °C.

Polymerization of norbornene using chiral metallocenes results in insoluble polymers exhibiting a glass transition temperature of about 210 °C. Although they have been shown by oligomerization to be tactic, no melting up to 500 °C

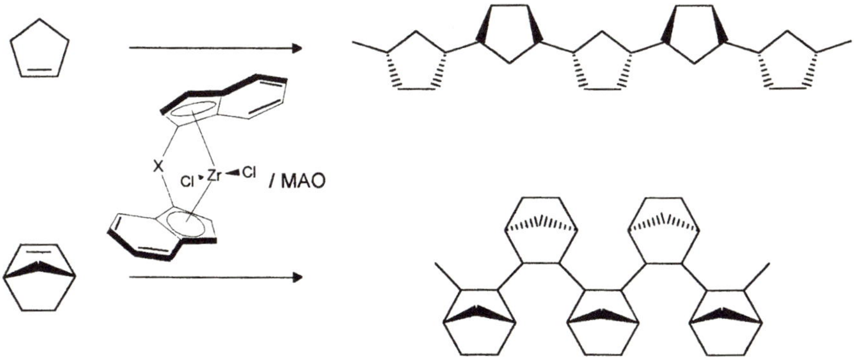

Fig. 19. Vinylpolymerization of cyclopentene and norbornene is possible using metallocene/MAO catalysts

Table 22. Polymerization of cyclopentene by metallocene/MAO catalysts. 45 ml toluene, 5 ml cyclopentene, [Zr] = 0.5 mmol/l, [MAO] = 100 mmol/l, 25 °C, 24 h

Metallocene	Activity (gPC/mmolZr·h)	%trans	M_n (g/mol)
[En(Ind)$_2$]ZrCl$_2$	6.0	<2	360
[En(IndH$_4$)$_2$]ZrCl$_2$	11.0	36	880
Cp$_2$ZrCl$_2$	5.0	3	560
[En(Cp)$_2$]ZrCl$_2$	4.0	<1	460
[En(3MeCp)$_2$]ZrCl$_2$	18.0	38	470
[En(3iPrCp)$_2$]ZrCl$_2$	5.0	66	1100
Ind$_2$ZrCl$_2$	0.7	27	800

has been observed. Wide angle X-ray scattering shows two amorphic halos, and for some samples additional sharp reflexes at small angles have been observed, so it is not clear if these polymers are crystalline [193].

5.2 Copolymers (COC)

While homopolymerization of cyclopentene results in 1,3-enchainment of the monomer units in copolymerisation, blocks of cyclic monomer units are rarely observed due to the unfavorable copolymerization parameters. The isolated cyclopentene units may be incorporated in a *cis*-1,2 or *cis*-1,3 fashion, their ratio depending on the catalyst used [194–196]. Thus ethene is able to compensate the steric hindrance at the α-carbon of the growing chain after the insertion of a cyclopentene.

The homopolymers of cycloolefins like norbornene or tetracyclododecene are not processable due to their high melting points and their insolubility in common organic solvents. By copolymerization of these cyclic olefins with ethene or α-olefins, cycloolefin copolymers (COC) are produced representing a new class of thermoplastic amorphous materials [197–203]. Early attempts to produce such copolymers were made using heterogenous TiCl$_4$/AlEt$_2$Cl catalysts. In the 1980s vanadium catalysts have been applied to these copolymerizations, but real progress was made utilizing metallocene catalysts for this purpose. They are about ten times more active than vanadium systems and, by choice of the metallocene, the comonomer distribution may be varied from statistical to alternating. Statistical copolymers are amorphous if more than 12–15 mol% of cycloolefin are incorporated in the polymer chain. The glass transition temperature can be varied over a wide range (Fig. 20) by choice of the cycloolefin and the amount of cycloolefin incorporated in the polymer chain.

In the case of ethene/norbornene copolymerization it is also possible to produce copolymers showing molecular weight distributions of Mw/Mn = 1.1

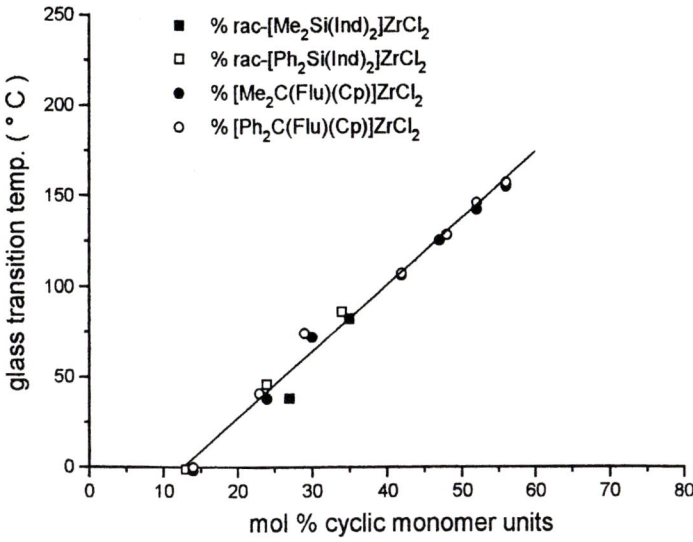

Fig. 20. Dependence of the glass transition temperature of ethene/norbornene copolymers produced by various metallocene/MAO catalysts on the incorporation of norbornene

Table 23. Properties of a random ethene norbornene copolymer containing 52 mol% of norbornene [105]

Mechanical properties	
Density (g/cm^3) 0	1.02
Glass transition temperature (°C)	150
e modulus, ISO 527 (MPa)	3100
Tensile strength, ISO 527 (MPa)	66
Elongation until break, ISO 527 (%)	2–3
Optical properties	
Clarity	white, clear
Anisotropy	very low
Refractive index	1.53

to 1.4 by controlling the polymerization conditions [204]. This "pseudo living polymerization" enables the production of block copolymers by changing the feed composition. Statistical copolymers are transparent due to their amorphous character, colorless and show a high optical anisotropy (Table 23). Due to their high carbon/hydrogen ratio these polymers have a high refractive index (1.53 for an ethene/norbornene copolymer at 50 mol% incorporation). Their stability against hydrolysis and chemical degradation in combination with their stiffness and very good processability makes them interesting materials for optical applications, for example in compact discs, lenses, and optical fibers [205].

In the case of alternating copolymers, beside the glass transition temperature (130 °C in case of ethene/norborene copolymerization), a melting point is observed (295 °C for a totally alternating copolymer). The melting point as well as the crystallinity of these copolymers may be influenced by choice of the metallocene and the conditions of polymerization. Compared to the statistical copolymers the alternating structures show a better resistance in resistance against unpolar solvents. The diameter of the crystallites is about 0.05–1 mm and thus these copolymers are transparent. Thermoplastic processing is possible at 300–330 °C. Similar alternating structures are obtained by ring opening polymerization of multicycle polyolefins followed by hydrogenation of the unsaturated polymer.

6 Other Monomers

6.1 Styrene

In 1985, the first pure syndiotactic polystyrene was synthesized by Idemitsu Kosan using transition metal compounds and methylaluminoxane as cocatalyst [80]. Since this time, a great number of patents for technical applications of syndiotactic polystyrene have been claimed [80, 206]. With a melting point of 270 °C, this material exhibits the highest melting point of most polymers made of a single monomer. Its properties range from good resistance against chemicals, water and steam to good rigidity and electronic properties.

Some half sandwich titanium compounds with cyclopentadienyl ligands have proven to be most active, but soluble tetraethyoxytitanium also shows a certain amount of activity. In contrast to olefin polymerization, titanocenes are more active than zirconocenes and fluoro ligands are better than chloro ligands. Table 24 [207] compares some catalysts for the polymerization of styrene.

Table 24. Synthesis of syndiotactic polystyrene using metallocene/MAO catalysts

Catalyst	Temp. (°C)	Activity (kgPS/molM·h)	T_m (°C)	M_w (kg/mol)	M_w/M_n
$CpTiCl_3$	10	109	267	390	3.6
	30	477	263	230	2.2
$(C_5Me_5)TiCl_3$	30	3.5	277	186	2.3
	50	15.4	275	169	3.6
$(C_5Me_5)ZrCl_3$	30	0.01	249	20	2.2
$(C_5Me_5)TiCl_3$	50	690	275	660	2.0
$CpTiF_3$	30	2400	261	380	1.8
	50	1700	257	100	2.0

While polymers, prepared with pentamethylcyclopentadienyl titanium dichloride show a high melting point of 277 °C, this is reduced to 275 °C for the fluorinated compound, although the latter has an activity which is 200 times higher. The molecular weights vary between 20 000 and 660 000.

Polymerization at various temperatures allows the calculation of the overall activation energy of polymerization to be between 20 and 40 kg/mol for the system Ti(OEt)$_4$/MAO. The catalysts are stable for at least 2 h. Activity which is low below 30 °C is enhanced exponentially at temperatures up to 70 °C. The monomer concentration has a major effect on the polymerization activity.

6.2 Dienes

Metallocenes polymerize non-conjugated and conjugated dienes. A new type of cyclopolymerization affording ring structures, interspaced by CH$_2$-groups of non-conjugated dienes, was observed by Waymouth and associates using 1,5-dienes [208–210]. A 1,2-insertion of the terminal double bond into the zirconium carbon bond is followed by an intramolecular cyclization forming a ring (Fig. 21) [211].

The microstructure and the properties depend on the *cis/trans* ratio of the ring bonding and on the stereochemistry between the rings. Poly(methylene-1,3-cyclopentane) obtained by cyclopolymerization of 1,5-hexadiene shows four different structures from which the *trans* isotactic structure is predominant, when using simple biscyclopentadienyl compounds. Higher substituted (pentamethyl) zirconocenes yield mainly *cis*-connected polymers which are highly crystalline and have melting points up to 190 °C.

Optically active ansa metallocenes (R-En(IndH)$_2$ZrCl$_2$) polymerize the diene to optically active trans-isotactic polymers while only optically active oligomers and inactive polymers are obtained with olefins [156].

Cyclopolymerizations of functionalized 1,6-dienes like 4-trimethylsilyloxy-1,6-heptadiene are also possible, using B(C$_6$F$_5$)$_3$ as cocatalyst and bis(pentamethylcyclopentadienyl)zirconocenes. After hydrolysis with HCl, polymethylene-3- hydroxycyclohexane is formed. With the same catalyst, Waymouth et al. [212] succeeded in polymerizing 5-*N,N*-diisopropyl-amino-1-pentene and 4-*tert*-butyldimethylsiloxy-1-pentene.

Another possibility in the functionalization of polyolefins is given by the copolymerization of olefins with borane monomers. Chung and Rhubrigut [213] and Chung [214] polymerize 5-hexenyl, 9-borabicyclo[3,3,1]nonane together with various α-olefins, such as ethene, propene, 1-butene using both heterogeneous catalysts (TiCl$_3$/AlEt$_2$Cl) and metallocene catalysts such as Cp$_2$ZrCl$_2$ and En(ind)$_2$ZrCl$_2$ with MAO as cocatalyst.

The homogeneous catalysts were much more effective. The high molecular weight polymers contain 1–2 mol% of the organoboroane units, which were reacted by ionic processes using NaOH/H$_2$O$_2$ reagents. The borane groups were completely converted to the corresponding hydroxyl groups. These

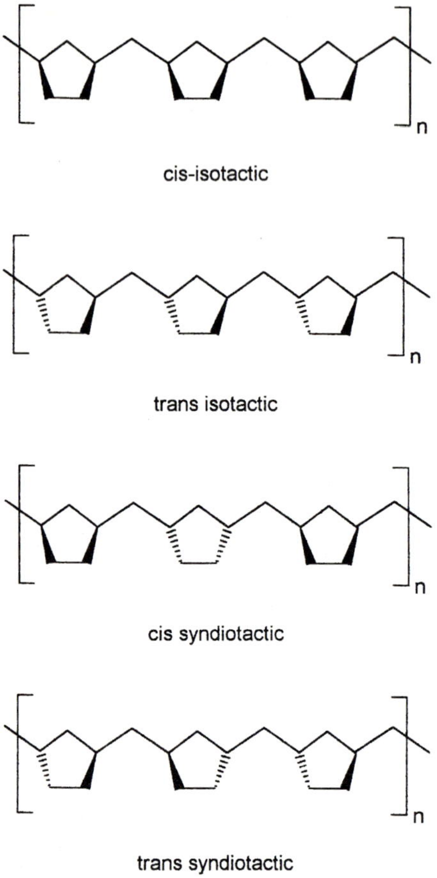

cis-isotactic

trans isotactic

cis syndiotactic

trans syndiotactic

Fig. 21. Possible microstructures of poly(methylenecyclopentane) which is produced by metallocene/MAO catalysts from 1,5 hexadiene

functional groups can be used, similar to vinyl end groups, to graft and block copolymers. Some interesting polymers, such as PE-PMMA, PE-polycaprolactone, PP-PMMA, and PP-polyvinyl acetate have been synthesized and have been used as agents in polymer blends [215].

Block copolymers of olefins and acrylates or vinylesters can be obtained by lanthanoid compounds. Living polyethylene-biscyclopentadienyl samarium systems can continue the polymerization with acrylate monomers by group transfer polymerization or cyclolactone monomers by ring opening polymerization [216, 217].

As found for the polymerization of styrene, $CpTiCl_3$/MAO and similar half-sandwich titanocenes are active catalysts for the polymerization of conjugated 1,3-dienes (Table 25) [218]. Butadiene, 1,3-pentadiene, 2-methyl-1,3-pentadiene and 2,3-dimethylbutadiene yield polymers with different

Table 25. Polymerization of dienes using CpTiCl$_3$/MAO at 20 °C, 2 ml monomer, 16 ml toluene, [Ti] = 10^{-5} mol/l, [MAO] = 10^{-2} mol/l

Monomer	Time (h)	Conv (%)	cis	1,2	trans	Tm (°)
Butadiene	0.75	42.9	80	18	2	–
(E)-1,3-Pentadiene	1	57.1	43	57	0	–
(Z)-1,3-Pentadiene	100	35	> 99	–	–	waxy
2,3-Dimethylbutadiene	1.2	85.3	> 99	–	–	120
(E)-2-Methylpentadiene	18	14.7	> 99	–	–	136
4-Methylpentadiene	0.1	100	–	99	–	95
(E, E)-2,4-Hexadiene	17	100	–	70	30	–

cis-1,4-, trans-1,4- and 1,2-structures depending on the polymerization temperature. A change in the stereospecificity was observed by Ricci et al. [219]. Polypentadiene with mainly cis-1,4-structures were obtained at + 20 °C while at − 20 °C a crystalline, 1,2-syndiotactic polymer was produced. This effect is attributed to a different mode of coordination of the monomer to the metallocene, which is mainly cis-η^4 at + 20 °C and trans η^2 at −20 °C.

6.3 Methylmethacrylate

The substitution of methylaluminoxane by other lewis acid cocatalysts like tetraphenylborates leads to catalysts which may tolerate functional groups in the monomer. Collins et al. investigated the polymerization of methylmethacrylate by two-component catalysts comprised of a metallocenedimethyl and a methylmetallocenium ion using tetraphenylborate as counterion [220, 221]. They obtained polymethylmethacrylate and in the case of a chiral metallocene [En(IndH4)$_2$]ZrCl$_2$ highly isotactic polymer was produced. Based on kinetic investigations a mechanism was proposed involving an intermolecular Michael addition of a neutral enolate with a complexed monomer. Based on the narrow molecular weight distribution of 1.2–1.5 the polymerization seems to having a living character. Deng et al. investigated the polymerization of methylmethacrylate initiated by chiral zirconocenedimethyl combined with [Ph$_3$C][B(C$_6$F$_5$)$_4$] and found highly isotactic polymers when an aluminium or zincalkyl is added as third component (Table 26) [222]. Investigation of the microstructure showed the catalysts to work by enantiomorphic site control.

Without the third component no activity at all was observed. They proposed a mechanism similar to the one given by Yasuda et al. [216, 217] for polymerization of methylmethacrylate by lanthanocenes which are isoelectronic with alkylzirconocenium ions. The role of the third component in this mechanism is not very clear. Nevertheless polymerization of polar monomers by metallocene catalysts is an open field of research and investigations are just beginning.

Table 26. Polymerization of methylmethacrylate using chiral zirconocenedimethyl complexes in combination with trityltetrakis(pentafluorphenyl) borate and zincalkyls

Metallocene	ZnR$_2$	Conv. (%)	mm	mr	rr	M$_n$ (kg/mol)	M$_w$/M$_n$
[En(Ind)$_2$]ZrMe$_2$	CH$_2$Me	33	96.5	2.5	1.0	393	1.43
	(CH$_2$)$_3$Me	82	94.5	3.5	2.0	600	1.30
	(CH$_2$)$_2$CH=CH$_2$	48	98.0	1.3	0.7	599	1.30
[En(IndH$_4$)]ZrMe$_2$	CH$_2$Me	79	94.4	3.7	1.9	320	1.32
	(CH$_2$)$_3$Me	68	95.1	3.7	1.2	405	1.30
	(CH$_2$)$_2$CH=CH$_2$	64	95.5	3.3	1.2	339	1.25
[Me$_2$Si(Ind)$_2$]ZrMe$_2$	CH$_2$Me	25	86.8	5.8	7.4	114	1.71
	(CH$_2$)$_3$Me	38	84.6	7.5	7.9	371	1.41
	(CH$_2$)$_2$CH=CH$_2$	34	91.4	5.3	2.3	345	1.34

7 Conclusion

Metallocenes are very versatile catalysts for the production of polyolefins, polystyrene and copolymers. Some polymers such as syndiotactic polypropene, syndiotactic polystyrene, cycloolefin copolymers, optically active oligomers, and polymethylenecycloalkenes can be produced only by metallocene catalysts. It is possible to tailor the microstructure of polymers by changing the ligand structure of the metallocene. The effect and influence of the ligands can more and more be predicted and understood by molecular modeling and other calculations.

It is estimated that by the year 2000 nearly every second new polyolefin plant will run with metallocene systems. A lot of polymer-producing companies try to use mainly supported metallocene catalysts in their running plants, the so-called "drop in" technology; others build plants specially equipped for these new types of catalysts.

The copolymerization of olefins together with polar monomers is just beginning and has a great potential.

A big step is also being made by metallocene research to understand the mechanism and the nature of the Ziegler-Natta catalysis.

8 References

1. Brintzinger HH, Fischer D, Mülhaupt R, Rieger B, Waymouth R (1995) Angew Chem Int Ed Eng 107: 1255
2. Thayer AM (1995) Chem. Eng. News
3. Clayfield TE (1995) Kunststoffe-Plast Europe 85: 9

4. Huang J, Rempel GL (1995) Prog Polym Sci 20: 459
5. Gupta VK, Satish S, Bhardwag IS (1994) Macromol Chem Phys C34: 439
6. Statistical specificiation
7. Ziegler K, Holzkamp E, Breil H, Martin H (1955) Angew Chem 67: 541
8. Natta G (1956) Angew Chem 68: 393
9. Woo TW, Fan L, Ziegler T (1955). In: Fink G, Mülhaupt R, Brintzinger HH (eds) Ziegler Catalysts. Springer, Berlin, p 291
10. Kaminsky W, Rabe O, Schauwienold AM, Schupfner GU, Hanss J, Kopf J (1955) J Organomet Chem 497: 181
11. Wilkinson G, Birmingham IM (1954) J Am Chem Soc 76: 4281
12. Keii T (1972) Kinetics of Ziegler-Natta Polymerization. Kodansha Ltd, Tokyo and Chapman & Hall Ltd, London
13. Boor J (1970) Ind Eng Chem Prod Res Dev 9: 437
14. Breslow DS, Newburg NR (1957) J Am Chem Soc 79: 5072
15. Natta G, Pino P, Corradini P, Danusso F, Mantica E, Mazzanti G, Moraglio G (1955) J Am Chem Soc 77: 1708
16. Belov GP, Kuznetsov VI, Solovyeva TI, Chirkov NM, Ivanchev SS (1970) Makromol Chem 140: 213
17. Dyachkovskii FS, Shilova AK, Shilov AE (1967) J Polym Sci 16: 2333
18. Patat F, Sinn H (1958) Angew Chem 70: 496
19. Chien JCW, Hsieh JTT (1975). In: Chien, JCW (ed) Coordination Polymerization, Academic Press, New York, p 305
20. Clauss K, Bestian H (1962) Justus Liebigs Ann Chem 654: 8
21. Henrici-Olivé G, Olivé S (1969) J Organomet Chem 16: 339
22. Reichert KH, Schoetter E (1968) Z Physi Chem 57: 74
23. Fink G (1972) Polym Prepr Am Chem Soc Div Polym Chem 13: 443
24. Reichert KH, Meyer KR (1973) Makromol Chem 169: 163
25. Long WP, Breslow DS (1975) Justus Liebigs Ann Chem, p 463
26. Andresen A, Cordes HG, Herwig J, Kaminsky K, Merck A, Mottweiler R, Pein J, Sinn H, Vollmer HJ (1976) Angew Chem Int Ed Engl 15: 630
27. Sinn H, Kaminsky W (1980) Adv Organomet Chem 18: 99
28. Wild FRWP, Zsolnai L, Huttner G, Brintzinger HH (1982) J Organomet Chem 232: 233
29. Ewen JA, Jones RL, Razavi A, Ferrara JP (1988) J Am Chem Soc 110: 6255
30. Kaminsky W, Miri M, Sinn H, Woldt R (1983) Makromol Chem Rapid Commun 4: 417
31. Kaminsky W, Steiger R (1988) Polyhedron 7: 2375
32. Tritto I, Li S, Sacchi MC, Zannoni G (1993) Macromolecules 26: 7112
33. Sinn H, Bliemeister J, Clausnitzer D, Tikwe L, Winter H, Zarncke O (1988). In: Kaminsky W, Sinn H (eds) Transition Metals and Organometallics as Catalysts for Olefin Polymerization, Springer, Berlin, p 257
34. Piccolrovazzi N, Pino P, Consiglio G, Sironi A, Moret M, Organometallics 9: 3098
35. Resconi L, Bossi S, Abis L (1990) Macromolecules 23: 4489
36. Cam D, Giannini U (1992) Makromol Chem 193: 1049
37. Siedle AR, Newmark RA, Lamanna WM, Schroepfer JN (1990) Polyhedron 9: 301
38. Mason MR, Smith JM, Bott SG, Barron AR (1993), J Am Chem Soc 115: 4971
39. Harlan CJ, Mason MR, Barron AR (1994) Organometallics 13: 2957
40. Sinn H (1995) Macromol Symp 97: 27
41. Barron AR (1995) Macromol Symp 97: 15
42. Eisch JJ, Pombrick SI, Zheng GX (1993) Organometallics 12: 3856
43. Gassmann PG, Callstrom MR (1987) J Am Chem Soc 109: 7875
44. Sishta C, Hathorn RM, Marks TJ (1992) J Am Chem Soc 114: 1112
45. Hlatky GG, Upton DJ, Turner HW (1990) US Pat Appl 459921; Chem Abstr 1991, 115, 256897 v
46. Yang X, Stern CL, Marks TJ (1991) Organometallics 10: 840
47. Zambelli A, Luongo P, Grassi A (1989) Macromolecules 22: 2186
48. Yang X, Stern CL, Marks TJ (1994) J Am Chem Soc 116: 10015
49. Chien JCW, Tsai WM, Rausch MD (1991) J Am Chem Soc 113: 8570
50. Bochmann M, Lancaster SJ (1993) Organometallics 12: 633
51. Bochmann M, Lancaster SL (1994) Angew Chem 106: 1715; Angew Chem Int Ed Engl 33: 1634
52. Kaminsky W, Bark A, Steiger R (1992) J Mol Catal 74: 109

53. Kaminsky W (1995) Macromol Symp 97: 79
54. Kaminsky W, Miri M, Sinn H, Woldt R (1983) Makromol Chem Rapid Commun 4: 417
55. Tait P (1988) in: Kaminsky W, Sinn H (eds) Transition Metals and Organo- metallics as Catalysts for Olefin Polymerization, Springer Press, Berlin, p 309
56. Chien JCW, Wang BP (1989) J Polym Sci Part A 27: 1539
57. Kaminsky W, Engehausen R, Zoumis K, Spaleck W, Rohrmann J (1992) Makromol Chem 193: 1643
58. Kaminsky W, Schlobohm M (1986) Macromol Symp 4: 103
59. Ahlers A, Kaminsky W (1988) Makromol Chem Rapid Commun 9: 457
60. Alt HG, Palackal SJ, Zenk R (1995) Dechema Jahrestagung I.6. Wiesbaden
61. Kaminsky W, Lüker H (1989) Makromol Chem Rapid Commun 5: 225
62. Herfert N, Fink G (1992) Makromol Chem 193: 1359
63. Tsutsui T, Kashiwa N (1988) Polym Commun 29: 180
64. Denger C, Haase U, Fink G (1991) Makromol Chem Rapid Commun 12: 697
65. Zambelli A, Grassi A, Galimberti M, Mazzochi R, Piemontesi F (1991) Makromol Chem Rapid Commun 12: 523
66. Kaminsky W, Drögemüller H (1989). In: Reichert KH, Geiseler W (eds) Polymer Reaction Engineering, VCH, Berlin, p 372
67. Chien JCW, He P (1991) J Polym Sci Polym Chem Ed 29: 1395
68. Heiland K, Kaminsky W (1992) Makromol Chem 193: 601
69. Allen RD (1983) J Elastom Plast 15: 19
70. Martuscelli E, Sticotti G, Massari P (1993) Polymer 34: 3671
71. Mirabella jr FM (1992) Polym Mater Sci Eng 67: 303
72. Kaminsky W, Miri M (1985) J Polym Sci Polym Chem Ed 23: 2151
73. Chien JCW, He D (1991) J Polym Sci Part A 29: 1585
74. Shapiro PJ, Bunnel E, Schaefer WP, Bercaw JE (1990) Organometallics 9: 867
75. Stevens JC, Timmers FJ, Wilson DR, Schmidt GF, Nickias PN, Rosen RK, Knight GW, Lay FY (1990) Eur Pat Appl, 416815
76. Stevens J (1993) Proc MetCon Houston May 26 –28, 157
77. Soga K, Park JR, Shiono T (1991) Polym Commun 10: 310
78. Pellecchia C, Proto A, Zambelli A (1992) Macromolecules 25: 4490
79. Aaltonen P, Seppälä J (1994) Eur Polym J 30: 683
80. JP 0267 328 (1990), Idemitsu Kosan Co. Ltd., inv.: K. Funski, M. Yamazaki, Chem Abstr 113, 39169 p
81. Busico V, Cipullo R, Corradini P (1993) Makomol Chem Rappid Commun 14: 97
82. Grassi A, Zambelli A, Resconi L, Albizzati E, Mazzocchi R (1989) Macromolecules 21: 617
83. Soga K, Shiono T, Takemura S, Kaminsky W (1987) Makromol Chem Rapid Commun 8: 305
84. Cheng HN, Ewen JA (1989) Makromol Chem 190: 1931
85. Rieger B, Mu X, Mallin DT, Rausch MD, Chien JCW (1990) Macromolecules 23: 3559
86. Schupfner G, Kaminsky W (1995) J Mol Catal A Chem 102: 59
87. Busico V, Cipullo R (1994) J Am Chem Soc 117: 1652
88. Sheldon RA, Fueno T, Tsuntsugu T, Kurukawa J (1965) J Polym Sci Part B 3: 23
89. Ewen JA (1984) J Am Chem Soc 106: 6355
90. Kaminsky W, Küpler K, Brintzinger HH, Wild FRWP (1985) Angew Chem 97: 507
91. Kaminsky W, Küpler K, Niedoba S (1986) Makromol Chem Macromol Symp 3: 377
92. Farina M, DiSilvestro G, Terragni A (1995) Makromol Chem Phys 196: 353
93. Resconi L, Piemontesi F, Franciscono G, Abis L, Fiorani T (1992) J Am Chem Soc 114: 1025
94. Resconi L, Abis L, Franciscono G (1992) Macromolecules 25: 6814
95. Kaminsky W, Arndt M (1994) in Soga K, Terano M: Catalysts Design for Tailor-made Polyolefins. Kodansha Ltd., Tokyo, p 179
96. Spaleck W, Antberg M, Aulbach M, Bachmann B, Dolle V, Haftka S, Küber F, Rohrmann J, Winter A (1994) in Fink G, Mülhaupt R, Brintzinger HH (eds.): Ziegler Catalysts. Springer Verlag, Berlin, p 83
97. Spaleck W, Küber F, Winter A, Rohrmann J, Bachmann B, Kiprof P, Behn J, Herrmann WA (1994) Organometallics 13: 954
98. Spaleck W, Aulbach M, Bachmann B, Küber F, Winter A (1995) Macromol Symp 89: 221
99. Jüngling S, Mülhaupt R, Stehling U, Brintzinger HH, Fischer D, Langhauser F (1995) Macromol Symp 97: 205

100. Resconi L, Fait A, Piemontesi F, Colonnesi M, Rychlicki H, Zeigler R (1995) Macromolecules 28: 6667
101. Spaleck W, Antberg A, Rohrmann J, Winter A, Bachmann B, Kiprof P, Behn J, Herrmann WA (1992) Angew Chem 104: 1373 Angew Chem Int Ed Engl 31: 1347
102. Stehling U, Diebold J, Kirsten R, Röll W, Brintzinger HH, Jüngling S, Mülhaupt R, Langhauser F (1994) Organometallics 13: 964
103. Mise T, Miya S, Yamazaki H, Chem Letters (1989), 1853.
104. Pino P, Cioni P, Wei J, Rotzinger B, Arizzi S (1988) in Quirk RP (ed.) Transition Metal Catalyzed Polymerizations. Cambridge University Press, Cambridge, p 1
105. Pino P, Cioni P, Wei J (1987) J Am Chem Soc 109: 6189
106. Waymouth R, Pino P (1990) J Am Chem Soc 112: 4911
107. Pino P, Galimberti M (1989) J Organomet Chem 370: 1
108. Corradini P, Guerra G (1988) in Transition Metal Catalyzed Polymerizations. Cambridge University Press, Cambridge, p 553
109. Corradini P, Busico V, Guerra, G (1987) in Kaminsky W, Sinn H: Transition Metals and Organometallics as Catalysts for Olefin Polymerisation. Springer Verlag, Berlin, p 337
110. Corradini P, Guerra G, Vacatello M, Villani V (1988) Gazz Chim Ital 118: 173
111. Venditto V, Guerra G, Corradini P, Fusco R (1991) Polymer 31: 530
112. Corradini P, Guerra G (1991) Progr Polym Sci 16: 239
113. Corradini P (1993) Makromol Chem Macromol Symp 66: 11
114. Ewen JA, Jones RL, Razavi A, Ferrara J (1988) J Am Chem Soc 110: 6255
115. Ewen JA, Elder MJ, Jones RL, Curtis S, Cheng HN (1991) in Keii T, Soga K (eds.) Catalytic Olefin Polymerization. Kodansha Ltd., Tokyo, p 439
116. Ewen JA, Elder MJ, Jones RL, Haspeslagh L, Atwood JL, Bott SG, Robinson K (1991) Polym Prepr Am Chem Soc 32: 469
117. Razavi A, Nafpliotis L, Vereecke D, DenDauw K, Atwood JL, Thewald U (1995) Macromol Symp 89: 345
118. Razavi A, Peters L, Nafpliotis L, Atwood JL (1994) in Soga K, Terano M (eds): Catalysts Design for Tailor-made Polyolefins. Kodansha Ltd., Tokyo
119. Farina M, DiSilvestro G, Sozzani P (1991) Progr Polym Sci 16: 219
120. Farina M, DiSilvestro G, Sozzani P (1993) Macromolecules 26: 946
121. Herfert N, Fink G (1993) Makromol Chem Macromol Symp 66: 157
122. Ewen JA (1995) Macromol Symp 89: 181
123. Spaleck W, Aulbach M, Bachmann B, Küber F, Winter A (1995) Macromol Symp 89: 237
124. Ewen JA, Elder MJ (1995) in Fink G, Mülhaupt R, Brintzinger HH (eds): Ziegler Catalysts. Springer Verlag, Berlin, p 99
125. Razavi A, Vereecke D, Peters L, DenDauw K, Nafpliotis L, Atwood JL in Fink G, Mülhaupt R, Brintzinger HH (eds): Ziegler Catalysts. Springer Verlag, Berlin, p 111
126. Mallin DT, Rausch MD, Lin GY, Dong S, Chien JCW (1990) J Am Chem Soc 112: 2030
127. Chien JCW, Llinas GH, Rausch MD, Lin GY, Winter HH (1991) J Am Chem Soc 113: 8569
128. Chien JCW, Llinas GH, Rausch MD, Lin GY, Winter HH, Atwood JL, Bott SG (1992) J Polym Sci Polym Chem 30: 2601
129. Llinas GH, Day RO, Rausch MD, Chien JCW (1993) Organometallics 12: 1283
130. Llin GY, Mallin DT, Chien JCW, Winter HH (1991) Macromolecules 24: 850
131. Llinas GH, Dong SH, Mallin DT, Rausch MD, Lin GY, Winter HH, Chien JCW (1992) Macromolecules 25: 1242
132. Gauthier WJ, Corrigan JF, Taylor NJ, Collins S (1995) Macromolecules 28: 3771
133. Gauthier WJ, Collins S (1995) Macromolecules 28: 3779
134. Erker G, Nolte R, Tsay YH, Krüger K (1989) Angew Chem 101: 642
135. Erker G (1992) Pure Appl Chem 64: 393
136. Erker G, Aulbach M, Wingbergmühle D, Krüger K, Werner S (1993) Chem Ber 126: 755
137. Erker G, Aulbach M, Knickmeier M, Wingbergmühle D, Krüger K, Nolte M, Werner S (1993) J Am Chem Soc 115: 4590
138. Erker G, Aulbach M, Krüger C, Werner S (1993) J Organomet Chem 450: 1
139. Coates GW, Waymouth RM (1995) Science 267: 217
140. Waymouth RM, Pino P (1990) J Am Chem Soc 112: 4911
141. Pino P, Galimberti M (1989) J Organomet Chem 370: 1
142. Kraudelat H, Brintzinger HH (1990) Angew Chem 102: 1459
143. Longo P, Proto A, Grassi A, Ammendola P (1991) Macromolecules 24: 4624

144. Busico V, Cipullo R (1995) Macromol Symp 89: 277
145. Busico V, Cipullo R (1994) J Am Chem Soc 116: 9329
146. Leclerc MK, Brintzinger HH (1995) J Am Chem Soc 117: 1652
147. Kaminsky W, Ahlers A, Möller-Lindenhof N (1989) Angew Chem 101: 1304
148. Kaminsky W, Ahlers A, Rabe O, König W (1993) in Enders D, Gais H, keim W: Organic Synthesis via Organometallics. Vieweg Verlag, Braunschweig, p 151
149. Kaminsky W (1995) Macromol Symp 89: 203
150. Silvestri R, Resconi L, Pelliconi A (1995) Proceedings of the International Congress on Metallocene Polymers Metallocene '95. Schotland Business research Inc., Brussels, p 207
151. Rieger B, Mu X, Mallin DT, Rausch MD, Chien JCW (1990) Macromolecules 23: 3559
152. Soga K, Shiono T, Takemura S, Kaminsky W (1987) Makromol Chem Rapid Commun 8: 305
153. Langhauser F, Fischer D, Seelert S (1995) Proceedings of the International Congress on Metallocene Polymers Metallocenes '95. Schotland Business Research Inc., Brussels p 243
154. Langius LJM (1995) Kunststoffe 85: 1122
155. Hungenberg KD, Kerth J, Langhauser F, Marczinke B, Schlund R (1995) Macromol Symp 89: 363
156. Antberg M, Dolle V, Haftka S, Rohrmann J, Spaleck W, Winter A, Zimmermann HJ (1991) Makromol. Chem. Macromol Symp 48/49: 333
157. Chowdhury J, Moore S (1993) Chem Eng 36
158. Kaminaka M, Soga K (1991) Makromol Chem Rapid Commun 12: 367
159. Soga K, Kaminaka M (1992) Makromol Chem Rapid Commun 13: 221
160. Soga K, Kaminaka M (1993) Makromol Chem 194: 1745
161. Soga K, Kaminaka M, Shiono T (1993) Proceedings of the Worldwide Metallocene Conference MetCon'93. Catalysts Consultants Inc., Houston
162. Soga K, Kim HJ, Shiono T (1994) Makromol Chem Rapid Commun 15: 139
163. Chien JCW, He D (1991) J Polym Sci Part A Polym Chem 29: 1603
164. Collins S, Kelly WM, Holden DA (1992) Macromolecules 25: 1780
165. Janiak C, Rieger B (1994) Angew Makromol Chem 215: 47
166. Lee D, Yoon K (1994) Macromol Rapid Commun 15: 841
167. Kaminsky W (1981) in Quirk RP (ed.) Transition Metal Catalyzed Polymerizations, Alkenes and Dienes.
168. Soga K, Arai T, Nozawa H, Uozumi T (1995) Macromol Symp 97: 53
169. Antberg M, Böhm L, Rohrmann J (1989) Eur Pat Appl 89122199.6 to Hoechst AG
170. Antberg M, Lüker H, Böhm L (1988) Eur Pat Appl 88108658.1 to Hoechst AG
171. Antberg M, Herrmann HF, Rohrmann J (1992) Eur Pat Appl 92100051.9 to Hoechst AG
172. Little IR, McNally JP (1993) Eur Pat Appl 93306666.4 to BP Chem Ltd
173. Chabrand CJ, McNally JP, Little IR (1993) Eur Pat Appl 93306665 to BP Chem Ltd
174. Marks TJ (1992) Acc Chem Res 25: 57
175. Kaminsky W, Renner F (1993) Makromol Chem Rapid Commun 14: 239
176. Kaminsky W, Renner F, Winkelbach H (1994) Proceedings of the MetCon '94. Houston TX, p
177. Sacchi MC, Zucchi D, Tritto I, Locatelli P (1995) Macromol Rapid Commun 16: 581
178. Calderon N (1972) J Macromol Sci Revs Macromol Sci C7: 105
179. Ivin KJ (1983): Olefin Metathesis. Academic Press, New York.
180. Novak BM, Risse W, Grubbs RH (1992) Adv Polym Sci 102: 47
181. Schultz RG (1966) Polym Letters 4: 451
182. Tanielian C, Kienneman A, Osparuch T (1979) Can J Chem 57: 2022
183. Sen A, Lai TW (1982) Organometallics 1: 415
184. Sen A, Lai TW, Thomas RR (1988) J Organomet Chem 359: 569
185. Mehler C, Risse W (1991) Makromol Chem Rapid Commun 12: 255
186. Mehler C, Risse W (1992) Makromol Chem Rapid Commun 13: 455
187. Breunig S, Risse W (1992) Makromol chem 193: 2915
188. Arndt M, Engehausen R, Kaminsky W, Zoumis K (1995) J Mol Catal A: Chem 101: 171
189. Arndt M, Kaminsky W (1995) Macromol Symp 97: 225
190. Collins S, Kelly WM (1992) Macromolecules 25: 233
191. Kelly WM, Taylor NJ, Collins S (1995) Macromolecules 27: 447
192. Arndt M, Kaminsky W (1995) Macromol Symp 95: 167
193. Arndt M, Grimm B, Kaminsky W, Zachmann HG (1995) Europhys Conf Abstr 19F: 42
194. Kaminsky W, Spiehl R (1989) Makromol Chem 190: 515
195. Arndt M (1994) PhD thesis University of Hamburg. Verlag Shaker, Aachen

196. Jershow A, Ernst E, Herrmann W, Müller N (1995) Macromolecules 28: 7095
197. Kaminsky W, Bark A, Arndt M (1991) Makromol Chem Macromol Symp 47: 83
198. Kaminsky W, Noll A (1993) Polym Bull 31: 175
199. Arndt M, Kaminsky W, Schupfner GU (1995) Proceedings of the International Congress on Metallocene Polymers Metallocenes '95. Schotland Business Research Inc., Brussels, p 403
200. Kaminsky W, Noll A (1955) in Fink G, Mülhaupt R, Brintzinger HH (eds): Ziegler Catalysts. Springer Verlag, Berlin, p 149
201. Benedikt GM, Goodall BL, Marchant NS, Rhodes LF (1994) New J Chem 18: 105
202. Benedikt GM, Goodall BL, Marchant NS, Rhodes LF (1994) Proceedings of the Worldwide Metallocene Conference MetCon'94. Catalyst Consultant Inc, Houston Texas
203. Kaminsky W, Engehausen R, Kopf J (1995) Angew Chem 107: 2469
204. Chedron H, Brekner MJ, Osan F (1994) Angew Makromol Chem 223: 121
205. Land HT (1995) Proceedings of the International Congress on Metallocene Polymers Metallocenes '95. Scotland Business Research Inc., Brussels, p 217
206. JP 0392 345 (1991), Idemitsu Kosan Co. Ltd., inv.: A Nakano, Chem Abstr 115, 51539h
207. Lenk S (1996) PhD thesis, University of Hamburg
208. Resconi L, Waymouth RM (1990) J Am Chem Soc 112: 4953
209. Resconi L, Coates GW, Mogstad A, Waymouth RM (1991) J Macromol Sci Chem Ed A28: 1255
210. Kesti MR, Waymouth RM (1991) J Am Chem Soc 114: 3565
211. Coates GW, Waymouth RM (1993) J Am Chem Soc 115: 91
212. Kesti MR, Coates GW, Waymouth RM (1992) J Am Chem Soc 114: 9679
213. Chung TC, Rhubright D (1993) Macromolecules 26: 3019
214. Chung TC (1995) Macromol Symp 89: 151
215 Rösch J, Mülhaupt R (1993) Makromol Chem Rapid Commun 14: 503
216. Yasuda H, Yamamoto H, Yokota K, Miyaka S, Nakamura (1992) J Am Chem Soc 114: 4908
217. Yasuda H, Yamamoto H, Yamashita K, Yokato K, Nakamura A, Miyada S (1993) Macromolecules 26: 7134
218. Porri L, Giarrusso A, Ricci G (1991) Prog Polym Sci 16: 405
219. Ricci G, Porri L, Giarrusso A (1995) Macromol Symp 89: 383
220. Collins S, Ward DG (1992) J Am Chem Soc 114: 5460
221. Collins S, Ward DG, Suddaby KH (1994) Macromolecules 27: 7222
222. Deng H, Shiono T, Soga K (1995) Macromolecules 28: 3067

Editor: H. Ringsdorf
Received: May 1996

Author Index Volumes 101-127

Author Index Vols. 1-100 see Vol. 100

Adolf, D. B. see Ediger, M. D..: Vol. 116, pp. 73-110.
Aharoni, S. M. and *Edwards, S. F.*: Rigid Polymer Networks. Vol. 118, pp. 1-231.
Améduri, B., Boutevin, B. and Gramain, P.: Synthesis of Block Copolymers by Radical Polymerization and Telomerization. Vol. 127, pp. 87-142.
Améduri, B. and *Boutevin, B.*: Synthesis and Properties of Fluorinated Telechelic Monodispersed Compounds. Vol. 102, pp. 133-170.
Amselem, S. see Domb, A. J.: Vol. 107, pp. 93-142.
Andreis, M. and *Koenig, J. L.*: Application of Nitrogen-15 NMR to Polymers. Vol. 124, pp. 191-238.
Angiolini, L. see Carlini, C.: Vol. 123, pp. 127-214.
Anseth, K. S., Newman, S. M. and *Bowman, C. N.*: Polymeric Dental Composites: Properties and Reaction Behavior of Multimethacrylate Dental Restorations. Vol. 122, pp. 177-218.
Armitage, B. A. see O'Brien, D. F.: Vol. 126, pp. 53-58.
Arndt, M. see Kaminski, W.: Vol. 127, pp. 143-187.
Arnold Jr., F. E. and *Arnold, F. E.*: Rigid-Rod Polymers and Molecular Composites. Vol. 117, pp. 257-296.
Arshady, R.: Polymer Synthesis via Activated Esters: A New Dimension of Creativity in Macromolecular Chemistry. Vol. 111, pp. 1-42.

Bahar, I., Erman, B. and *Monnerie, L.*: Effect of Molecular Structure on Local Chain Dynamics: Analytical Approaches and Computational Methods. Vol. 116, pp. 145-206.
Baltá-Calleja, F. J., González Arche, A., Ezquerra, T. A., Santa Cruz, C., Batallón, F., Frick, B. and *López Cabarcos, E.*: Structure and Properties of Ferroelectric Copolymers of Poly(vinylidene) Fluoride. Vol. 108, pp. 1-48.
Barshtein, G. R. and *Sabsai, O. Y.*: Compositions with Mineralorganic Fillers. Vol. 101, pp.1-28.
Batallán, F. see Baltá-Calleja, F. J.: Vol. 108, pp. 1-48.
Barton, J. see Hunkeler, D.: Vol. 112, pp. 115-134.
Bell, C. L. and *Peppas, N. A.*: Biomedical Membranes from Hydrogels and Interpolymer Complexes. Vol. 122, pp. 125-176.
Bennett, D. E. see O'Brien, D. F.: Vol. 126, pp. 53-84.
Berry, G.C.: Static and Dynamic Light Scattering on Moderately Concentraded Solutions: Isotropic Solutions of Flexible and Rodlike Chains and Nematic Solutions of Rodlike Chains. Vol. 114, pp. 233-290.
Bershtein, V. A. and *Ryzhov, V. A.*: Far Infrared Spectroscopy of Polymers. Vol. 114, pp. 43-122.
Bigg, D. M.: Thermal Conductivity of Heterophase Polymer Compositions. Vol. 119, pp. 1-30.
Binder, K.: Phase Transitions in Polymer Blends and Block Copolymer Melts: Some Recent

Developments. Vol. 112, pp. 115-134.
Bird, R. B. see Curtiss, C. F.: Vol. 125, pp. 1-102.
Biswas, M. and *Mukherjee, A.*: Synthesis and Evaluation of Metal-Containing Polymers. Vol. 115, pp. 89-124.
Boutevin, B. and *Robin, J. J.*: Synthesis and Properties of Fluorinated Diols. Vol. 102. pp. 105-132.
Boutevin, B. see Amédouri, B.: Vol. 102, pp. 133-170.
Boutevin, B. see Améduri, B.: Vol. 127, pp. 87-142.
Bowman, C. N. see Anseth, K. S.: Vol. 122, pp. 177-218.
Boyd, R. H.: Prediction of Polymer Crystal Structures and Properties. Vol. 116, pp. 1-26.
Bronnikov, S. V., *Vettegren, V. I.* and Frenkel, S. Y.: Kinetics of Deformation and Relaxation in Highly Oriented Polymers. Vol. 125, pp. 103-146.
Bruza, K. J. see Kirchhoff, R. A.: Vol. 117, pp. 1-66.
Burban, J. H. see Cussler, E. L.: Vol. 110, pp. 67-80.

Cameron, N. R. and *Sherrington, D. C.*: High Internal Phase Emulsions (HIPEs)-Structure, Properties and Use in Polymer Preparation. Vol. 126, pp. 163-214.
Candau, F. see Hunkeler, D.: Vol. 112, pp. 115-134.
Capek, I.: Kinetics of the Free-Radical Emulsion Polymerization of Vinyl Chloride. Vol. 120, pp. 135-206.
Carlini, C. and *Angiolini, L.*: Polymers as Free Radical Photoinitiators. Vol. 123, pp. 127-214.
Casas-Vazquez, J. see Jou, D.: Vol. 120, pp. 207-266.
Chen, P. see Jaffe, M.: Vol. 117, pp. 297-328.
Choe, E.-W. see Jaffe, M.: Vol. 117, pp. 297-328.
Chow, T. S.: Glassy State Relaxation and Deformation in Polymers. Vol. 103, pp. 149-190.
Chung, T.-S. see Jaffe, M.: Vol. 117, pp. 297-328.
Connell, J. W. see Hergenrother, P. M.: Vol. 117, pp. 67-110.
Criado-Sancho, M. see Jou, D.: Vol. 120, pp. 207-266.
Curro, J.G. see Schweizer, K.S.: Vol. 116, pp. 319-378.
Curtiss, C. F. and *Bird, R. B.*: Statistical Mechanics of Transport Phenomena: Polymeric Liquid Mixtures. Vol. 125, pp. 1-102.
Cussler, E. L., *Wang, K. L.* and *Burban, J. H.*: Hydrogels as Separation Agents. Vol. 110, pp. 67-80.

Dimonie, M. V. see Hunkeler, D.: Vol. 112, pp. 115-134.
Dodd, L. R. and *Theodorou, D. N.*: Atomistic Monte Carlo Simulation and Continuum Mean Field Theory of the Structure and Equation of State Properties of Alkane and Polymer Melts. Vol. 116, pp. 249-282.
Doelker, E.: Cellulose Derivatives. Vol. 107, pp. 199-266.
Domb, A. J., *Amselem, S.*, *Shah, J.* and *Maniar, M.*: Polyanhydrides: Synthesis and Characterization. Vol.107, pp. 93-142.
Dubrovskii, S. A. see Kazanskii, K. S.: Vol. 104, pp. 97-134.
Dunkin, I. R. see Steinke, J.: Vol. 123, pp. 81-126.

Economy, J. and *Goranov, K.*: Thermotropic Liquid Crystalline Polymers for High Performance Applications. Vol. 117, pp. 221-256.
Ediger M. D. and *Adolf, D. B.*: Brownian Dynamics Simulations of Local Polymer Dynamics. Vol. 116, pp. 73-110.

Edwards, S. F. see Aharoni, S. M.: Vol. 118, pp. 1-231.
Endo, T. see Yagci, Y.: Vol. 127, pp. 59-86.
Erman, B. see Bahar, I.: Vol. 116, pp. 145-206.
Ezquerra, T. A. see Baltá-Calleja, F. J.: Vol. 108, pp. 1-48.

Fendler, J.H.: Membrane-Mimetic Approach to Advanced Materials. Vol. 113, pp. 1-209.
Fetters, L. J. see Xu, Z.: Vol. 120, pp. 1-50.
Förster, S. and *Schmidt, M.*: Polyelectrolytes in Solution. Vol. 120, pp. 51-134.
Frenkel, S. Y. see Bronnikov, S. V.: Vol. 125, pp. 103-146.
Frick, B. see Baltá-Calleja, F. J.: Vol. 108, pp. 1-48.
Fridman, M. L.: see Terent'eva, J. P.: Vol. 101, pp. 29-64.

Ganesh, K. see Kishore, K.: Vol. 121, pp. 81-122.
Geckeler, K. E. see Rivas, B.: Vol. 102, pp. 171-188.
Geckeler, K. E.: Soluble Polymer Supports for Liquid-Phase Synthesis. Vol. 121, pp. 31-80.
Gehrke, S. H.: Synthesis, Equilibrium Swelling, Kinetics Permeability and Applications of Environmentally Responsive Gels. Vol. 110, pp. 81-144.
Godovsky, D. Y.: Electron Behavior and Magnetic Properties Polymer-Nanocomposites. Vol. 119, pp. 79-122.
González Arche, A. see Baltá-Calleja, F. J.: Vol. 108, pp. 1-48.
Goranov, K. see Economy, J.: Vol. 117, pp. 221-256.
Gramain, P. see Améduri, B.: Vol. 127, pp. 87-142.
Grosberg, A. and *Nechaev, S.*: Polymer Topology. Vol. 106, pp. 1-30.
Grubbs, R., Risse, W. and *Novac, B.*: The Development of Well-defined Catalysts for Ring-Opening Olefin Metathesis. Vol. 102, pp. 47-72.
van Gunsteren, W. F. see Gusev, A. A.: Vol. 116, pp. 207-248.
Gusev, A. A., Müller-Plathe, F., van Gunsteren, W. F. and *Suter, U. W.*: Dynamics of Small Molecules in Bulk Polymers. Vol. 116, pp. 207-248.
Guillot, J. see Hunkeler, D.: Vol. 112, pp. 115-134.
Guyot, A. and *Tauer, K.*: Reactive Surfactants in Emulsion Polymerization. Vol. 111, pp. 43-66.

Hadjichristidis, N. see Xu, Z.: Vol. 120, pp. 1-50.
Hall, H. K. see Penelle, J.: Vol. 102, pp. 73-104.
Hammouda, B.: SANS from Homogeneous Polymer Mixtures: A Unified Overview. Vol. 106, pp. 87-134.
Hedrick, J. L. see Hergenrother, P. M.: Vol. 117, pp. 67-110.
Heller, J.: Poly (Ortho Esters). Vol. 107, pp. 41-92.
Hemielec, A. A. see Hunkeler, D.: Vol. 112, pp. 115-134.
Hergenrother, P. M., Connell, J. W., Labadie, J. W. and *Hedrick, J. L.*: Poly(arylene ether)s Containing Heterocyclic Units. Vol. 117, pp. 67-110.
Hiramatsu, N. see Matsushige, M.: Vol. 125, pp. 147-186.
Hirasa, O. see Suzuki, M.: Vol. 110, pp. 241-262.
Hirotsu, S.: Coexistence of Phases and the Nature of First-Order Transition in Poly-N-isopropylacrylamide Gels. Vol. 110, pp. 1-26.
Hunkeler, D., Candau, F., Pichot, C., Hemielec, A. E., Xie, T. Y., Barton, J., Vaskova, V., Guillot, J., Dimonie, M. V., Reichert, K. H.: Heterophase Polymerization: A Physical and Kinetic Comparision and Categorization. Vol. 112, pp. 115-134.

Ichikawa, T. see Yoshida, H.: Vol. 105, pp. 3-36.
Ilavsky, M.: Effect on Phase Transition on Swelling and Mechanical Behavior of Synthetic Hydrogels. Vol. 109, pp. 173-206.
Inomata, H. see Saito, S.: Vol. 106, pp. 207-232.
Irie, M.: Stimuli-Responsive Poly(N-isopropylacrylamide), Photo- and Chemical-Induced Phase Transitions. Vol. 110, pp. 49-66.
Ise, N. see Matsuoka, H.: Vol. 114, pp. 187-232.
Ivanov, A. E. see Zubov, V. P.: Vol. 104, pp. 135-176.

Jaffe, M., Chen, P., Choe, E.-W., Chung, T.-S. and *Makhija, S.*: High Performance Polymer Blends. Vol. 117, pp. 297-328.
Jou, D., Casas-Vazquez, J. and *Criado-Sancho, M.*: Thermodynamics of Polymer Solutions under Flow: Phase Separation and Polymer Degradation. Vol. 120, pp. 207-266.

Kaetsu, I.: Radiation Synthesis of Polymeric Materials for Biomedical and Biochemical Applications. Vol. 105, pp. 81-98.
Kaminski, W. and *Arndt, M.*: Metallocenes for Polymer Catalysis. Vol. 127, pp. 143-187.
Kammer, H. W., Kressler, H. and *Kummerloewe, C.*: Phase Behavior of Polymer Blends - Effects of Thermodynamics and Rheology. Vol. 106, pp. 31-86.
Kandyrin, L. B. and *Kuleznev, V. N.*: The Dependence of Viscosity on the Composition of Concentrated Dispersions and the Free Volume Concept of Disperse Systems. Vol. 103, pp. 103-148.
Kaneko, M. see Ramaraj, R.: Vol. 123, pp. 215-242.
Kang, E. T., Neoh, K. G. and *Tan, K. L.*: X-Ray Photoelectron Spectroscopic Studies of Electroactive Polymers. Vol. 106, pp. 135-190.
Kazanskii, K. S. and *Dubrovskii, S. A.*: Chemistry and Physics of „Agricultural" Hydrogels. Vol. 104, pp. 97-134.
Kennedy, J. P. see Majoros, I.: Vol. 112, pp. 1-113.
Khokhlov, A., Starodybtzev, S. and *Vasilevskaya, V.*: Conformational Transitions of Polymer Gels: Theory and Experiment. Vol. 109, pp. 121-172.
Kilian, H. G. and *Pieper, T.*: Packing of Chain Segments. A Method for Describing X-Ray Patterns of Crystalline, Liquid Crystalline and Non-Crystalline Polymers. Vol. 108, pp. 49-90.
Kishore, K. and *Ganesh, K.*: Polymers Containing Disulfide, Tetrasulfide, Diselenide and Ditelluride Linkages in the Main Chain. Vol. 121, pp. 81-122.
Klier, J. see Scranton, A. B.: Vol. 122, pp. 1-54.
Kobayashi, S., Shoda, S. and *Uyama, H.*: Enzymatic Polymerization and Oligomerization. Vol. 121, pp. 1-30.
Koenig, J. L. see Andreis, M.: Vol. 124, pp. 191-238.
Kokufuta, E.: Novel Applications for Stimulus-Sensitive Polymer Gels in the Preparation of Functional Immobilized Biocatalysts. Vol. 110, pp. 157-178.
Konno, M. see Saito, S.: Vol. 109, pp. 207-232.
Kopecek, J. see Putnam, D.: Vol. 122, pp. 55-124.
Kressler, J. see Kammer, H. W.: Vol. 106, pp. 31-86.
Kirchhoff, R. A. and *Bruza, K. J.*: Polymers from Benzocyclobutenes. Vol. 117, pp. 1-66.
Kuleznev, V. N. see Kandyrin, L. B.: Vol. 103, pp. 103-148.
Kulichkhin, S. G. see Malkin, A. Y.: Vol. 101, pp. 217-258.
Kuchanov, S. I.: Modern Aspects of Quantitative Theory of Free-Radical Copolymerization.

Vol. 103, pp. 1-102.
Kummerloewe, C. see Kammer, H. W.: Vol. 106, pp. 31-86.
Kuznetsova, N. P. see Samsonov, G. V.: Vol. 104, pp. 1-50.

Labadie, J. W. see Hergenrother, P. M.: Vol. 117, pp. 67-110.
Lamparski, H. G. see O'Brien, D. F.: Vol. 126, pp. 53-84.
Laschewsky, A.: Molecular Concepts, Self-Organisation and Properties of Polysoaps. Vol. 124, pp. 1-86.
Laso, M. see Leontidis, E.: Vol. 116, pp. 283-318.
Lazár, M. and *Rychlý, R.*: Oxidation of Hydrocarbon Polymers. Vol. 102, pp. 189-222.
Lenz, R. W.: Biodegradable Polymers. Vol. 107, pp. 1-40.
Leontidis, E., de Pablo, J. J., Laso, M. and *Suter, U. W.*: A Critical Evaluation of Novel Algorithms for the Off-Lattice Monte Carlo Simulation of Condensed Polymer Phases. Vol. 116, pp. 283-318.
Lesec, J. see Viovy, J.-L.: Vol. 114, pp. 1-42.
Liang, G. L. see Sumpter, B. G.: Vol. 116, pp. 27-72.
Lin, J. and *Sherrington, D. C.*: Recent Developments in the Synthesis, Thermostability and Liquid Crystal Properties of Aromatic Polyamides. Vol. 111, pp. 177-220.
López Cabarcos, E. see Baltá-Calleja, F. J.: Vol. 108, pp. 1-48.

Majoros, I., Nagy, A. and *Kennedy, J. P.*: Conventional and Living Carbocationic Polymerizations United. I. A Comprehensive Model and New Diagnostic Method to Probe the Mechanism of Homopolymerizations. Vol. 112, pp. 1-113.
Makhija, S. see Jaffe, M.: Vol. 117, pp. 297-328.
Malkin, A. Y. and *Kulichkhin, S. G.*: Rheokinetics of Curing. Vol. 101, pp. 217-258.
Maniar, M. see Domb, A. J.: Vol. 107, pp. 93-142.
Matsumoto, A.: Free-Radical Crosslinking Polymerization and Copolymerization of Multivinyl Compounds. Vol. 123, pp. 41-80.
Matsuoka, H. and *Ise, N.*: Small-Angle and Ultra-Small Angle Scattering Study of the Ordered Structure in Polyelectrolyte Solutions and Colloidal Dispersions. Vol. 114, pp. 187-232.
Matsushige, K., Hiramatsu, N. and *Okabe, H.*: Ultrasonic Spectroscopy for Polymeric Materials. Vol. 125, pp. 147-186.
Mays, W. see Xu, Z.: Vol. 120, pp. 1-50.
Mikos, A. G. see Thomson, R. C.: Vol. 122, pp. 245-274.
Miyasaka, K.: PVA-Iodine Complexes: Formation, Structure and Properties. Vol. 108. pp. 91-130.
Monnerie, L. see Bahar, I.: Vol. 116, pp. 145-206.
Morishima, Y.: Photoinduced Electron Transfer in Amphiphilic Polyelectrolyte Systems. Vol. 104, pp. 51-96.
Müllen, K. see Scherf, U.: Vol. 123, pp. 1-40.
Müller-Plathe, F. see Gusev, A. A.: Vol. 116, pp. 207-248.
Mukerherjee, A. see Biswas, M.: Vol. 115, pp. 89-124.
Mylnikov, V.: Photoconducting Polymers. Vol. 115, pp. 1-88.

Nagy, A. see Majoros, I.: Vol. 112, pp. 1-113.
Nechaev, S. see Grosberg, A.: Vol. 106, pp. 1-30.

Neoh, K. G. see Kang, E. T.: Vol. 106, pp. 135-190.
Newman, S. M. see Anseth, K. S.: Vol. 122, pp. 177-218.
Noid, D. W. see Sumpter, B. G.: Vol. 116, pp. 27-72.
Novac, B. see Grubbs, R.: Vol. 102, pp. 47-72.
Novikov, V. V. see Privalko, V. P.: Vol. 119, pp. 31-78.

O'Brien, D. F., Armitage, B. A., Bennett, D. E. and *Lamparski, H. G.* : Polymerization and Domain Formation in Lipid Assemblies. Vol. 126, pp. 53-84
Ogasawara, M.: Application of Pulse Radiolysis to the Study of Polymers and Polymerizations. Vol.105, pp.37-80.
Okabe, H. see Matsushige, K.: Vol. 125, pp. 147-186.
Okada, M.: Ring-Opening Polymerization of Bicyclic and Spiro Compounds. Reactivities and Polymerization Mechanisms. Vol. 102, pp. 1-46.
Okano, T.: Molecular Design of Temperature-Responsive Polymers as Intelligent Materials. Vol. 110, pp. 179-198.
Onuki, A.: Theory of Phase Transition in Polymer Gels. Vol. 109, pp. 63-120.
Osad'ko, I.S.: Selective Spectroscopy of Chromophore Doped Polymers and Glasses. Vol. 114, pp. 123-186.

de Pablo, J. J. see Leontidis, E.: Vol. 116, pp. 283-318.
Padias, A. B. see Penelle, J.: Vol. 102, pp. 73-104.
Penelle, J., Hall, H. K., Padias, A. B. and *Tanaka, H.*: Captodative Olefins in Polymer Chemistry. Vol. 102, pp. 73-104.
Peppas, N. A. see Bell, C. L.: Vol. 122, pp. 125-176.
Pichot, C. see Hunkeler, D.: Vol. 112, pp. 115-134.
Pieper, T. see Kilian, H. G.: Vol. 108, pp. 49-90.
Pospíšil, J.: Functionalized Oligomers and Polymers as Stabilizers for Conventional Polymers. Vol. 101, pp. 65-168.
Pospíšil, J.: Aromatic and Heterocyclic Amines in Polymer Stabilization. Vol. 124, pp. 87-190.
Priddy, D. B.: Recent Advances in Styrene Polymerization. Vol. 111, pp. 67-114.
Priddy, D. B.: Thermal Discoloration Chemistry of Styrene-co-Acrylonitrile. Vol. 121, pp. 123-154.
Privalko, V. P. and *Novikov, V. V.*: Model Treatments of the Heat Conductivity of Heterogeneous Polymers. Vol. 119, pp 31-78.
Putnam, D. and *Kopecek, J.*: Polymer Conjugates with Anticancer Acitivity. Vol. 122, pp. 55-124.

Ramaraj, R. and *Kaneko, M.*: Metal Complex in Polymer Membrane as a Model for Photosynthetic Oxygen Evolving Center. Vol. 123, pp. 215-242.
Rangarajan, B. see Scranton, A. B.: Vol. 122, pp. 1-54.
Reichert, K. H. see Hunkeler, D.: Vol. 112, pp. 115-134.
Risse, W. see Grubbs, R.: Vol. 102, pp. 47-72.
Rivas, B. L. and *Geckeler, K. E.*: Synthesis and Metal Complexation of Poly(ethyleneimine) and Derivatives. Vol. 102, pp. 171-188.
Robin, J. J. see Boutevin, B.: Vol. 102, pp. 105-132.
Roe, R.-J.: MD Simulation Study of Glass Transition and Short Time Dynamics in Polymer Liquids. Vol. 116, pp. 111-114.

Ruckenstein, E.: Concentrated Emulsion Polymerization. Vol. 127, pp. 1-58.
Rusanov, A. L.: Novel Bis (Naphtalic Anhydrides) and Their Polyheteroarylenes with Improved Processability. Vol. 111, pp. 115-176.
Rychlý, J. see Lazár, M.: Vol. 102, pp. 189-222.
Ryzhov, V. A. see Bershtein, V. A.: Vol. 114, pp. 43-122.

Sabsai, O. Y. see Barshtein, G. R.: Vol. 101, pp. 1-28.
Saburov, V. V. see Zubov, V. P.: Vol. 104, pp. 135-176.
Saito, S., Konno, M. and *Inomata, H.*: Volume Phase Transition of N-Alkylacrylamide Gels. Vol. 109, pp. 207-232.
Samsonov, G. V. and *Kuznetsova, N. P.*: Crosslinked Polyelectrolytes in Biology. Vol. 104, pp. 1-50.
Santa Cruz, C. see Baltá-Calleja, F. J.: Vol. 108, pp. 1-48.
Sato, T. and *Teramoto, A.*: Concentrated Solutions of Liquid-Christalline Polymers. Vol. 126, pp. 85-162.
Scherf, U. and *Müllen, K.*: The Synthesis of Ladder Polymers. Vol. 123, pp. 1-40.
Schmidt, M. see Förster, S.: Vol. 120, pp. 51-134.
Schweizer, K. S.: Prism Theory of the Structure, Thermodynamics, and Phase Transitions of Polymer Liquids and Alloys. Vol. 116, pp. 319-378.
Scranton, A. B., Rangarajan, B. and *Klier, J.*: Biomedical Applications of Polyelectrolytes. Vol. 122, pp. 1-54.
Sefton, M. V. and *Stevenson, W. T. K.*: Microencapsulation of Live Animal Cells Using Polycrylates. Vol. 107, pp. 143-198.
Shamanin, V. V.: Bases of the Axiomatic Theory of Addition Polymerization. Vol. 112, pp. 135-180.
Sherrington, D. C. see Cameron, N. R., Vol. 126, pp. 163-214.
Sherrington, D. C. see Lin, J.: Vol. 111, pp. 177-220.
Sherrington, D. C. see Steinke, J.: Vol. 123, pp. 81-126.
Shibayama, M. see Tanaka, T.: Vol. 109, pp. 1-62.
Shoda, S. see Kobayashi, S.: Vol. 121, pp. 1-30.
Siegel, R. A.: Hydrophobic Weak Polyelectrolyte Gels: Studies of Swelling Equilibria and Kinetics. Vol. 109, pp. 233-268.
Singh, R. P. see Sivaram, S.: Vol. 101, pp. 169-216.
Sivaram, S. and *Singh, R. P.*: Degradation and Stabilization of Ethylene-Propylene Copolymers and Their Blends: A Critical Review. Vol. 101, pp. 169-216.
Starodybtzev, S. see Khokhlov, A.: Vol. 109, pp. 121-172.
Steinke, J., Sherrington, D. C. and *Dunkin, I. R.*: Imprinting of Synthetic Polymers Using Molecular Templates. Vol. 123, pp. 81-126.
Stenzenberger, H. D.: Addition Polyimides. Vol. 117, pp. 165-220.
Stevenson, W. T. K. see Sefton, M. V.: Vol. 107, pp. 143-198.
Sumpter, B. G., Noid, D. W., Liang, G. L. and *Wunderlich, B.*: Atomistic Dynamics of Macromolecular Crystals. Vol. 116, pp. 27-72.
Suter, U. W. see Gusev, A. A.: Vol. 116, pp. 207-248.
Suter, U. W. see Leontidis, E.: Vol. 116, pp. 283-318.
Suzuki, A.: Phase Transition in Gels of Sub-Millimeter Size Induced by Interaction with Stimuli. Vol. 110, pp. 199-240.
Suzuki, A. and *Hirasa, O.*: An Approach to Artifical Muscle by Polymer Gels due to Micro-Phase Separation. Vol. 110, pp. 241-262.

Tagawa, S.: Radiation Effects on Ion Beams on Polymers. Vol. 105, pp. 99-116.
Tan, K. L. see Kang, E. T.: Vol. 106, pp. 135-190.
Tanaka, T. see Penelle, J.: Vol. 102, pp. 73-104.
Tanaka, H. and *Shibayama, M.*: Phase Transition and Related Phenomena of Polymer Gels. Vol. 109, pp. 1-62.
Tauer, K. see Guyot, A.: Vol. 111, pp. 43-66.
Teramoto, A. see Sato, T.: Vol. 126, pp. 85-162.
Terent'eva, J. P. and *Fridman, M. L.*: Compositions Based on Aminoresins. Vol. 101, pp. 29-64.
Theodorou, D. N. see Dodd, L. R.: Vol. 116, pp. 249-282.
Thomson, R. C., Wake, M. C., Yaszemski, M. J. and *Mikos, A. G.*: Biodegradable Polymer Scaffolds to Regenerate Organs. Vol. 122, pp. 245-274.
Tokita, M.: Friction Between Polymer Networks of Gels and Solvent. Vol. 110, pp. 27-48.
Tsuruta, T.: Contemporary Topics in Polymeric Materials for Biomedical Applications. Vol. 126, pp. 1-52.

Uyama, H. see Kobayashi, S. : Vol. 121, pp. 1-30.

Vasilevskaya, V. see Khokhlov, A., Vol. 109, pp. 121-172.
Vaskova, V. see Hunkeler, D.: Vol. 112, pp. 115-134.
Verdugo, P.: Polymer Gel Phase Transition in Condensation-Decondensation of Secretory Products. Vol. 110, pp. 145-156.
Vettegren, V. I.: see Bronnikov, S. V.: Vol. 125, pp. 103-146.
Viovy, J.-L. and *Lesec, J.*: Separation of Macromolecules in Gels· Permeation Chromatography and Electrophoresis. Vol. 114, pp. 1-42.
Volksen, W.: Condensation Polyimides: Synthesis, Solution Behavior, and Imidization Characteristics. Vol. 117, pp. 111-164.

Wake, M. C. see Thomson, R. C.: Vol. 122, pp. 245-274.
Wang, K. L. see Cussler, E. L.: Vol. 110, pp. 67-80.
Wunderlich, B. see Sumpter, B. G.: Vol. 116, pp. 27-72.

Xie, T. Y. see Hunkeler, D.: Vol. 112, pp. 115-134.
Xu, Z., Hadjichristidis, N., Fetters, L. J. and *Mays, J. W.*: Structure/Chain-Flexibility Relationships of Polymers. Vol. 120, pp. 1-50.

Yagci, Y. and *Endo, T.*: N-Benzyl and N-Alkoxy Pyridium Salts as Thermal and Photochemical Initiators for Cationic Polymerization. Vol. 127, pp. 59-86.
Yannas, I. V.: Tissue Regeneration Templates Based on Collagen-Glycosaminoglycan Copolymers. Vol. 122, pp. 219-244.
Yamaoka, H.: Polymer Materials for Fusion Reactors. Vol. 105, pp. 117-144.
Yaszemski, M. J. see Thomson, R. C.: Vol. 122, pp. 245-274.
Yoshida, H. and *Ichikawa, T.*: Electron Spin Studies of Free Radicals in Irradiated Polymers. Vol. 105, pp. 3-36.

Zubov, V. P., Ivanov, A. E. and *Saburov, V. V.*: Polymer-Coated Adsorbents for the Separation of Biopolymers and Particles. Vol. 104, pp. 135-176.

Subject Index

Acrylamide 23, 30, 41
Acrylate 100, 102, 104, 113, 127, 131–133
–, acrylamide 95
–, acrylic acid 133
–, alkyl acrylate 127, 130
–, butyl acrylate 119, 132
–, ethyl acrylate 100, 106, 113, 127, 133
–, fluoroacrylate 109
–, methyl acrylate 100, 112–113, 128
Aging reactions 149–150
–, deactivation 150
–, methan production 148–149
AIBN 96, 131
Alkoxyamine 98, 102
Aluminiumalkyls 147–148
–, triethylaluminiumal 147
–, triisobuthylaluminiumal 148
–, trimethylaluminiumal 148
Aluminoxane 147–152
–, ethylaluminoxane 147
–, isobutylaluminoxane 152
–, methylaluminoxane 147–152
–, tert.-butylaluminoxane 147
– by ^{13}C-NMR 147
– by x-ray analysis 148
Anisotropy 177
Applications
– of hybrid fluorosilicones 118
–, curing 120
–, electronics 116
–, surface agent 109
–, textile treatment 109
–, transporation 116

Arkopal-N15 7
Azobisisobutyronitrile 18

Bistelomerization, see telomerization
Block copolymerization 80
Block copolymers 90, 93, 95, 98, 99, 105, 111, 117, 120–122, 127
–, amphiphilic 95, 120
–, diblock copolymers, fluorinated 116
–, – from multiinitiators 96–97
–, –, hard-b-soft 96
–, –, PBA-b-PS 119
–, –, PBuMA-b-PS 122, 123
–, –, PBzMA-b-PS 123
–, –, PDMS-b-PAN 121
–, –, PDMS-b-polydiethyl fumarate 121
–, –, PDMS-b-polymaleic anhydride 121
–, –, PEA-b-PMA 100
–, –, PEA-b-PS 106
–, –, PEO-b-PDMS 111
–, –, PMA-b-PEA 100
–, –, PMA-b-PEO 112
–, –, PMA-b-PS 100
–, –, PMMA-b-PBuA 102
–, –, PMMA-b-PCL 105
–, –, PMMA-b-polymethacrylic acid 112
–, –, PMMA-b-PS 104, 105
–, –, PMMA-b-PS 123, 125
–, –, PMMA-b-PV Ac 101
–, –, PÒMS-ty-b-PMMA 120
–, –, PPO-b-PFA 109
–, –, PS-b-P (acrylic acid) 100

–, –, PS-b-P acrylate   100
–, –, PS-b-PAN   110
–, –, PS-b-PBd   96, 100, 111
–, –, PS-b-PBuA   132
–, –, PS-b-PCL   105
–, –, PS-b-PCMSty   101
–, –, PS-b-PDMS   111
–, –, PS-b-PHEMA   105
–, –, PS-b-PIp   101
–, –, PS-b-PMA   113
–, –, PS-b-PMMA   100, 102, 105, 107, 108, 125, 133
–, –, PS-b-PVOH   129
–, –, PV Ac-b-PMMA   101
–, –, PV Ac-b-PS   101
–, multiblock copolymers   92, 95, 96, 120–122
–, triblock copolymers   113
–, –, amphyphilic   96
–, –, from multiinitiator   95
–, –, PEA-b-PMA-b-PMMA   100
–, –, PEG-b-PS-b-PEG   96
–, –, PEO-b-PDMS-b-POE   111
–, –, PMMA-b-PDMS-b-PMMA   112
–, –, PMMA-b-PS-b-PMMA   93, 94
–, –, PS-b-PBd-b-PS   101, 110
–, –, PS-b-PMMA-b-PS   93
Blockcopolymer   180
BMA   45
Bulk polymerization   21
Butadiene   96, 100, 128
Butyl methacrylate (BMA)   33, 44
Butylacrylate   11

Catalyst see metal, complex
Cationic active site   148
Cationic polymerization   59–86
– by charge transfer complexes   78
– by free radical promoted   71
–, photoinitiators for –   68
–, photosensitized –   77
–, redox initiated –   79
Cleavage of bond   113
–, carbon-carbon   113, 119–121, 123
–, carbon-halogen   113

–, –, C-Br   113
–, –, C-Cl   113, 114
–, –, C-I   113, 114
–, carbon-sulfur   113, 123
–, Silicon-silicon   113
–, sulfur-sulfur   113, 125
Coalescence of the droplets   9
$Co(CO)_4^-$ 53
COC cycloolefin copolymers   176–178
Collodial pathway to polymer composites   37
Comonomers   153, 154
Compatibilizer   110
Complex   104
–, cobalt oxime   104
–, FeCl3/benzoin   107
–, organoborane   104
Concentrated emulsion   4, 5, 21, 27
–, pathway to toughened polymeric latexes   41
–, polymerization   26
–, polymerization pathway to hydrophobic and hydrophilic microsponge molecular reservoirs   50
Conductive composite polymers   55
Controlled   100, 103, 126–130, see also polymerization
–, architectured polymers   90
Copolymer   92, 106
–, block, see block polymers
–, grafted, see grafted polymers
Copolymerization   153–156
–, parameters   154, 155
–, copolymers   154–156
– of styrene and methacrylic acid in concentrated emulsions   26
Cotelomerization, see telomerization
Counter radical, see radical
Counterions   148
Coupling reaction   106, 112
– by amidification   110
– by condensation   109
– by esterification   110
– by hydrosilylation   110, 111

Subject Index

– of telomers   106, 110, 111
Cross-linked polystyrene   52
Crystallinity   154
Cycles   90

Decane   11, 23, 24, 30
Density of polyethylene   154
Diblock, see block copolymers
2,5-Dihydro-2-oxo-3-hydroxy-4-phenylfuran   53
Divinylbenzene (DVB)   53

Elastomers
–, fluoroelastomers   116
–, Thermoplastic   110, 122
Elongation at the break point   43
Enantiomeric excess   165, 166
Encapsulation   49
–, of solid particles by the concentrated emulsion polymerization method   49
End group, see functional group
Ethene/propene elastomers   156, 170
Ethyl acrylate   11
(±) 2-Ethyl hexylacrylate   33
Ethyl methacrylate   11, 33
Ethyl methacrylate/water   16
Ethylbenzene   11
Ethylidenenorbornene   156
Extractables   154

Free radical polymerization   80
Functional group   92, 110, 112, 132
–, transesterification   112
–, trityl   119
Functionalization   117, 118
Functionalized polymers   180

Grafted copolymers   90, 99
–, PBd-g-PMA   102
–, PS-g-PS   102, 103
Grafting onto polystyrene   68

Half-sandwich complexes   150, 156
Hardness of polypropene   168–170
Heterogenization   171–174

– by recation on silica   172
– of metallocenes   171–174
Highly cross-linked hydrophobic polystyrene particles   52
HLB   24
Hydration forces   7
Hydrophilic-hydrophobic polymer composites   29
Hydrophilic-lipophilic balance (HLB)   13
Hydrophobic monomer   9

Immiscible polymers   110
Improved concentrated emulsion polymerization pathway   31
Inifer   125
Iniferter   98, 125, 126
–, Thiuram disulfide   125, 128, 130, 131, 133, 134
–, transfer constant   129
–, Xanthogen disulfide   110, 125, 128
Initer   122
Initiation, see also polymerization mechanism
–, electronic radiation   91
–, photochemical   91, 94
–, –, UV   94, 120, 127
–, –, visible light   94
–, radical   108
–, redox   108
–, thermal   91, 94, 117, 118
–, thermolysis   122
Initiator   102, 119, 125
–, AIBN, see AIBN
–, ammonium persulfate   116
–, amphiphilic   95
–, azo   92, 93
–, –, AIBN, see AIBN
–, –, alkylcarbonylzao   92
–, –, arylazooxyl   102, 135
–, –, azoperoxide   92
–, –, azoperoxyester   92, 93
–, –, diaromatic azo   124
–, –, polyazoderivatives   92

–, branched perfluoroalkane 119, 123
–, decomposition 92
–, functional 92, 94
–, macromolecular 95, 98
–, ω-trityl polymers 124
–, perester 93, 94
–, peroxide 92, 115
–, –, azoperoxyde 92
–, –, borane peroxide 104
–, –, diacyl peroxide 92
–, –, dialkyl peroxide 92
–, –, dicumyl peroxide 101
–, –, oligoperoxide 92
–, –, poly(THF)peroxide 98
–, –, polyperoxide 92
–, –, sucinic acid peroxide 98
–, phenyl radical 122
–, photoinitiator 134, see photochemical polymerization
–, polyamide 95
–, polyaromatic 95
–, polybutadiene 95
–, polycarbonate 95
–, polydiemthylsiloxanne 95
–, polyester 95
–, polyethylene glycol 95
–, polymeric 93, 95
–, polymethylmethacrylate 95
–, polystyrene 95
–, polyvinyl acetate 95
–, polyvinyl chloride 95
–, self 100
–, telechelic 104
–, tetraphenylethane 119
–, tetraphenylethane 120
Interfacial free energy 10
Interfacial tension 10
Intrinsic viscosities 43
Iodine transfer polymerization 114, 115
– of fluoromonomers 116
Irregularities 167
Isoprene, polymerization 100, 127
Isotacticity 161, 162

Kinetics 128
–, decomposition of initiator 92
–, dissociation constant 93
–, polymerization rate constants 119
–, transfer constant, see telogen

Lanthanocenes 181
Lightly cross-linked hydrophilic polyacrylamide particles 52
Lightly cross-linked polyacrylamide nutshells encapsulating loosely entangled poly (EO) molecules 52
Living polymerization 103, 118, 119, 122, 126, 127–133, 135, 137
–, anionic 90
–, cationic 98
–, –, poly(THF) 98
–, PMMA 104
Long chain branching 156

Macroinitiator 127, see initiator
–, macroperoxide 96
–, polyazomacroinitiator 95–97
–, –, PDMS 121
–, –, PMMA 120
–, –, poly(butyl azelate) 95
–, –, poly(butyl sebacate) 95
–, –, polyamide 95
–, –, polybutadiene 95
–, –, polycarbonate 95
–, –, polyether 95
–, –, polyethylacrylate 100
–, –, polyisoprenyl 97
–, –, polystyrol 97
Macromonomer 107
– of vinyl chloride 107
Macroradical 91, 98, 99, 100, 105, 127, 133, 137
–, control of 105
–, methylacrylate 100
–, recombination, see polymerization mechanism
MBSB composites 44

Mechanism
–, activation   172
–, chain-end control   160
–, enantiomorphic site control   158, 181
–, isomerization   166
–, á-hydrogen transfer   156, 160
Metal
–, acetoacetato   104, 135
–, complex, redox   116
–, metallic complex, see complex
–, salts of transition meetal   94
Metallocene   145–152, 156–160, 176–178
–, C_1-symmetry   163
–, C_2-symmetry   160
–, C_s-symmetry   161, 162
–, hafnocenes   152
–, half sandwich   178
–, historical development   145, 146
–, structures   150–152
–, titanocenes   145–152
–, zirconocenes   148–152
Methacrylate   100, 102
–, benzyl methacrylate   123
–, butyl methacrylate   95, 122, 123
–, hydroxy ethyl methacrylate(HEMA)   105
–, methyl methacrylate   100–108, 112, 120, 127, 128, 132–135
Methacrylic acid   26, 27
Methyl cellulose   16
Methyl methacrylate (MMA)   11, 33, 41, 44
Methylaluminoxine   147, 148
Molecular weight   26, 43, 99
–, distribution   167, 174, 176
Monomer, see polymerization
–, reactivity   123, 128

N-Alkoxy pyridinium salts   68–86
–, cationic polymerization by, -   69
–, free radical polymerization by, -   80
–, synthesis   69

N-Benzyl pyridinium salts   62–68
–, cationic polymerization by, -   62
–, synthesis   62
n-butyl methacrylate   11
n-Butyl methacrylate/water   16
Network   90
Nickel   104
Nitroxyl   98, 99, see also counter radical
Norbornene   174–176
Nutshells of highly cross-linked polystyrene encapsulating sparse poly (VBC) matrixes   52

Oil water (o/w) emulsion   9
Oligomerization   165–167
Oligomers
–, cooligomers, amphiphilic   112
–, –, PFA-b-PS-b-PFA   112
–, –, PHEA-b-PS-b-PHEA   112
–, –, photocrosslinkable   107
–, –, PS-b-PMMA   108
–, –, triblock   109
Ostwald ripening   9

Pentafluorophenyl-borate   148
Phase behavior and stability of concentrated emulsions of hydrocarbons in water   5
2-Phenyl-2-(Ò-styryoxyl)-ethanol   53
Photodissociation, see polymerization (photochemical)
Plastic waste
–, recovering   90
–, recycling   90
Polyacrylamide   25, 26, 41
– (dispersed phase)-polystyrene (continuous phase) composite   29
Polycondensation   90, 108
Polycycloolefins   174–178
–, copolymers   176–178
–, polycycloorbornene   175
–, polycyclopentene   175, 176
Polydiene   179, 180
–, cis-trans-microstructures   181

–, cyclopolymerization 179
–, cyclopolymerization 180
Polydispersity 99, 101, 102
Polyethylene 149–155
Polyethylene glycol 96
–, copolymers 153–156
–, linear low density (LIDPE) 153
–, long chain branching 156
–, molecular weight 152, 153
Polymer conversion 25, 27
Polymer latexes 9, 18
Polymer substrates for the immobilization of enzymes and cells 56
Polymer supported catalytic groups 55
Polymerization 109, 119, see block copolymers
– of a hydrophobic monomer 18, 23
– of styrene and acrylonitrile 109
–, controlled radical 103, 108, 113, 115, 118, 119, 123, 129, 135–137
–, –, ATRP 113
–, –, butadiene 128
–, –, butyl acrylate 118, 119
–, –, isobutyl vinyl ether 114
–, –, isopropene 100
–, –, methyl acrylate 100, 130
–, –, MMA 100, 101, 104, 113, 114, 118, 119, 124, 128, 134
–, –, p-chloromethyl styrene 103
–, –, styrene 100, 101, 113, 114, 118, 124, 128, 130–133
–, –, styrene fulfonic acid sodiium 101
–, –, vinyl acetate 101, 114, 128
–, coordinative 90
–, inoic 108, 109, 137
–, –, aninoic 90, 97
–, –, catinoic 90, 97, 105
–, mechanism, initiation 91, 116
–, –, propagation 91
–, –, termination 91, 98, 125
–, –, –, disproportionation 92, 98, 132
–, –, –, recombination 92, 98, 132

–, –, –, reversible 98–100, 105, 113, 115, 120, 122, 125, 127
–, –, transfer 125, see also telogen
–, photochemical 99, 119–121, 124, 126–134, see initiation (photochemical)
Polymethylmethacrylate 181
Polypropylene 160–170
–, atactic 167
–, isoblock 161, 162
–, isotactic 167–169
–, stereoblock 161, 162
–, syndiotactic 168, 169
Poly(styrenesulfonic acid) salt latex particles 37
Poly(vinylidine chloride)/Poly(butyl methacrylate) composites prepared via the concentrated emulsion Pathway 45
Polymethyl methacrylate 41
Polystyrene 21, 30, 178
– (dispersed phase)-polyacrylamide (continuous phase) composite 29
– latexes 37
Properties of polypropenes 167–170
–, crystallinity 170
–, density 168
–, hardness 168–170
–, light transmission 169
–, melt index 168, 171
–, melting point 167
–, modulus 168, 169
Polytriethylvinylbenzyl ammonium chloride (PEVAC) 53
Properties, elasticity 90
–, hydrophilicity 129
–, hydrophobic 129
–, impact resistance 90
–, mechanical 90, 122
–, refractive index 116
–, softness 129
–, surface 90
–, thermostability 116, 122
–, viscosity 90
Pseudo phase behavior 9

Subject Index

Radical 91, 92, 121
–, counter radical 91, 98, 99, 104, 135–137
–, –, indoline oxyl 100
–, –, nitroxyl 99–101, 134, 135, 137
–, –, PDMS diradical 121
–, –, persistant 99, 135, 137
–, –, phenyl 124
–, –, stable 98, 104, 135
–, –, triphenyl verdazyl 102
–, –, trityl 121, 122
–, dithiocarbamyl 132
–, fluorinated 123
–, radical trap 122
Reaction rates 150
Recovering 90
Recycling 90
Regiospecificity 157
–, 1,2-insertion 157
–, 2,1-insertion 157
Repulsive double layer forces 7
$Rh(CO)_2I_2^-$ 53
ROMP (ring opening polymerization) 175
Rubber toughened polystyrene composites 42

SBS 44
Seletive composite membranes 55
Sodium dodecyl sulfate (SDS) 5, 18, 30
Solution polymerization 25, 26
Sorbitan monooleate 30
Span 20 12, 14
Span 40 14
Span 80 12–14, 23, 24
Span 85 14
Stability of concentrated emulsions
– containing monomers 9
Stability of emulsions 9
Stabilization, β-agostic 166
Stepwise
–, block copolymer 117
–, polymerization 107, 119
–, telomerization 114, 117, 124

–, – of ethyl acrylate 107
–, – of fluorinated monomers 116–118
–, – of styrene 107
Stereospecificity 157–166
–, atactic 159
–, isotactic 159–162
–, pentads 158, 161
–, stereo- and isoblock 159
–, syndiotactic 159,
–, syndiotactic 163–166
Stress-strain curves 46
Styrene (ST) 11, 18, 26, 27, 30, 33, 44, 96, 103, 104, 127, 128, 130–133
–, Ò-methyl styrene 120
–, p-chloromethylstyrene 101
–, polymerization of 96, 100–102, 104–108, 110, 111, 119, 122–125, 127, 128
Styrene-butadiene-styrene three-block copolymer (SBS) 42
Styrene/water 16
Supporting of metallocenes 171–174
– on aluminia 172
– on silica 172, 173
– on starch 172
– on zeolithes 172

Tacticity 171
Telechelic 90, 113, 117, 119, 120, 125–129
–, aromatic 123
–, diisocyanate PS 112
–, dithiol polymers 109, 128
–, –, PBd 109
–, fluorinated 118
–, PDMS 111, 112
–, poly (L-lactide) 111
–, poly ethylene oxide 111
–, polyether diol 112
–, polystyrene 130
–, polysulfone 111
Telogen 113, 115, 116, 119, 126, 137

–, brominated   117
–, CBr$_4$ 106, 107
–, CCl$_4$ 107, 108, 113, 114
–, –, transfer   107
–, chemical change of   106, 119
–, chlorofluorotelogen   107
–, cleavage of telogen, see cleavage of bond
–, disulfide   126
–, –, transfer constant   126
–, mercaptan   109, 112
–, mercaptoacetic acid   109, 110
–, mercaptoethanol   111
–, Ò,ω-diiodoperfluoroalkane   116–118
–, perfluoroalkyl halide   115–117, 123
–, telechelic   106
–, tetraphenylethane   119, 120, 137
–, thiol, see mercaptan
–, thiuram, see iniferter
–, trityl mercaptan   124
–, –, transfer constant   124, 125
–, xanthogen, see iniferter
–, block cotelomer   105, 107, 111, 118, 124
–, fluorinated   116–118
Telomerization   114, 124, 128, 137
– of monomers
–, –, allyl acetate   114
–, –, butadiene   106, 110
–, –, ethyl acrylate   107
–, –, fluoromonomers   116–118
–, –, isoprene   106
–, –, methyl acrylate   106
–, –, nonconjugated diene   114
–, –, styrene   107
–, –, vinylidene chloride   114
–, bistelomerization   106–108
–, cotelomerization   116
–, dormant   105
–, living   105, 114, 116, 124
–, pseudoliving   105, 116

–, redox   113
Tempo   100, 101
Tensile strength   43
1,1,2,2-Tetrachloroethane   24, 25
Tetraphenylborate   148
Thermally latent initiators   62
Thermoplastic
–, elastomer   110, 122
–, fluroelastomer   116
–, PS-b-PDMS   111
Titanocenes   145–152
Toluene   24
Transfer
–, chain transfer   110, 122
–, degenerative transfer   114, 118, 119, 137
–, transfer constant, see telogen
Transfer agent, see teologen
Transparency   167, 168
Triblock, see block copolymers
Triton X-100   5
Tween 20   14
Tween 40   14
Tween 60   14
Two-step collodial pathway to polymer composites   37
Two-step method   32

UV radiation   94

VDC   45
Vinyl acetate   11
Vinylbenzyl chloride (VBC)   53
Vinylidene chloride (VDC)   45
Viscosity   7, 10

Water in oil (w/o) emulsion   9

Xanthogen disulfide   110, 125

Ziegler-Natta catalysis   145–148, 165, 167, 182

Springer-Verlag and the Environment

We at Springer-Verlag firmly believe that an international science publisher has a special obligation to the environment, and our corporate policies consistently reflect this conviction.

We also expect our business partners – paper mills, printers, packaging manufacturers, etc. – to commit themselves to using environmentally friendly materials and production processes.

The paper in this book is made from low- or no-chlorine pulp and is acid free, in conformance with international standards for paper permanency.

Printing: Saladruck, Berlin
Binding: Buchbinderei Lüderitz & Bauer, Berlin

CHEMISTRY LIBRARY
Hildebrand H

T
1 Month